Fundamentals of FOOD PROCESSING & PRESERVATION

THE EDITORS

Dr. Anjum Ayoub did his Ph.D. in Food Science and Technology from Sher-e-Kashmir University of Agricultural Sciences and Technology Jammu (J&K).M.tech in Food Processing & Technology from Gautam Buddha University Greater Noida (U.P) and B.tech in Food Technology from Islamic University of Science and Technology Pulwama Kashmir (J&K). Dr. Anjum Ayoub received CSIR-Senior Research Fellowship during Ph.D. and many several fellowships during his school days as well as in higher studies. He has published several research articles including book chapters in reputed national as well as in international journals and books respectively.

Dr. Fozia Hameed did her Ph.D. in Food Science and Technology from Sher-e-Kashmir University of Agricultural Sciences and Technology Jammu (J&K).M.Tech Food Technology (Food safety and Standards) from Sam Higginbottom Institute of Agriculture, Technology and Sciences(U.P) and B.tech in Food Technology from Islamic University of Science and Technology Pulwama Kashmir (J&K). Dr. Fozia Hameed received CSIR-Senior Research Fellowship during Ph.D. and many several fellowships during her school days as well as in higher studies. She has published several research articles including book chapters in reputed national as well as in international journals and books respectively.

Ms. Nadira Anjum is pursuing her Ph.D. in Food Science and Technology from Sher-e-Kashmir University of Agricultural Sciences and Technology Jammu (J&K).She has done M.Tech in Food Technology (Food process engineering) from Sam Higginbottom Institute of Agriculture, Technology and Sciences (U.P) and B.tech in Food Technology from Islamic University of Science and Technology Pulwama Kashmir (J&K).Ms.NadiraAnjum is receiving ICMR-Senior Research Fellowship in her Ph.D. also she has received several fellowships during her school days as well as in higher studies. She has published several research articles including book chapters in reputed national as well as in international journals and books respectively.

Fundamentals of FOOD PROCESSING & PRESERVATION

– EDITORS –

Dr. Anjum Ayoub

B.tech, M.Tech , Ph.D.

Dr. Fozia Hameed

B.tech, M.Tech , Ph.D.

Ms. Nadira Anjum

B.tech, M.Tech , Ph.D.

2022

Daya Publishing House®

A Division of

Astral International Pvt. Ltd.

New Delhi – 110 002

ISBN: 978-93-5461-167-4 (HB)

Published by : **Daya Publishing House®**
A Division of
Astral International Pvt. Ltd.
– ISO 9001:2015 Certified Company –
4736/23, Ansari Road, Darya Ganj
New Delhi-110 002
Ph. 011-43549197, 23278134
E-mail: info@astralint.com
Website: www.astralint.com

Digitally Printed at : **Replika Press Pvt. Ltd.**

PROFESSOR ANURADHA MISHRA
DEPARTMENT OF APPLIED CHEMISTRY
FORMER DEAN SCHOOL OF VOC. STUDIES & APPLIED SCIENCES
FORMER DEAN ACADEMICS
GAUTAM BUDHA UNIVERSITY
GREATER NOIDA - 201312

Foreword

Food is essentially a perishable product. It starts deteriorating from the moment it is harvested, collected, and stored. Food essentially needs to be stored and preserved often for prolonged periods of time. Food Processing is a very important science that apart from many other challenges, also deals with food preservation. For the welfare and good nutrition of the society, Food Processing must be given the attention it deserves especially in the current situation. The status of the food-processing sector varies by country and is influenced by many different factors. Since food is a fundamental human requirement and a continuous supply of fresh food cannot always be guaranteed, every country and every society needs food processing services in variable measure and form. The degree, to which refined foods find their way into the family's daily meal, has an impact on the food-processing industry's economic dimension. Food additives are not fresh, and their overuse in recent years has raised some serious health concerns. It has also put some visible blame on food technologists and on the government.

Food Science and Technology, according to one school of thought, are not viewed as two separate subjects topics, but are actually two large divisions of what is in fact, just one continuum. It's absolutely clear without even the slightest doubt, that there is a subtle and nuanced relationship between Food Science and Food Technology. Food Science and Technology can provide opportunities for the convergence of various types of knowledge bases for the benefit of society. To study the extensibility and extrusion properties of bread dough, physicists, mathematicians, and rheologists are brought together; microbiologists, nutritionists, and toxicologists are brought together to investigate the safety of a processed food item; microwave engineers, packaging engineers, and statisticians are brought together to define a quality controlled high speed food processing unit; and political scientists are brought together to define a quality controlled high speed unit process.

Mr. Anjum Ayoub, Ms. Fozia Hameed, and Ms. Nadira Anjum, Research Scholars in the Division of Food Science and Technology, Sher-e-Kashmir University of Agricultural Sciences and Technology, Jammu, have written this textbook on "Fundamentals of Food Processing and Preservation". I congratulate Mr. Anjum Ayoub, Ms. Fozia Hameed, and Ms. Nadira Anjum on their book. Their book covers a wide range of topics including food processing principles and applications; conventional and innovative food preservation technologies; food bio-preservation; use of enzymes for food preservation; basics and applied concepts of edible packing for food preservation; recent trends in active, smart and intelligent packaging for food products; food freezing technologies; use of biotechnology in food processing; hurdle technology; nanotechnology in food processing, challenges, and opportunity; and a very logically derived conclusion. All chapters of the book are comprised of fundamental information.

This textbook is expected to be extremely useful to a wide variety of scholars and professional pertaining to many varied streams of science and technology. Postgraduate students and research scholars of Food Science & Technology, Nutrition & Food Science, Home Economists; Catering & Nutrition Schools: Quality Nutrition Breeding Studies, in particular, will benefit from the wealth of knowledge provided in the book and excellent reference content.

(Anuradha Mishra)

PREFACE

The transformation of raw materials into food or feed, or from one type of food to another, is referred to as food processing. Food processing transforms clean, harvested crops or butchered animal products into appealing, marketable food or feed products with a longer shelf life. Food processing dates back to ancient times, when it included sun drying, salt preservation, and various forms of cooking (such as roasting, smoking, steaming, and oven baking). In the nineteenth and twentieth centuries, modern food processing technology was primarily designed to meet military requirements. Nicolas Appert invented a vacuum bottling technique in 1809, followed by Peter Durand's development of canning in 1810, and Louis Pasteur's pasteurization in 1862, which was a major advancement in ensuring the microbiological protection of food.With advancements such as spray drying, juice concentrates, freeze drying, and the introduction of artificial sweeteners, colouring agents, and preservatives, the growing consumer culture has aided the growth of the food processing sector. Products such as dried instant soups, reconstituted fruits and juices, and instant/quick cooking meals were introduced as a result of these innovations.

Toxin reduction, preservation, ease of marketing and delivery tasks, and increased food quality are all benefits of modern food processing techniques. Furthermore, it increases the seasonal availability of many foods, allows for long-distance transportation of delicate perishable foods, and makes many types of foods safe to consume by preventing spoilage and pathogenic microorganisms. Processed foods are less likely to spoil quickly than fresh foods, and they are ideally suited to long-distance transport from the source to the customer. Processed foods help to reduce food shortages and increase population wellbeing because they are easily accessible to the general public. People with allergies, diabetics, and those who are unable to eat those traditional food components benefit from modern food processing. Extra nutrients, such as vitamins and minerals, may be added to food during processing. Food processing is a burgeoning industry in India. It is a critical tool for farmers to reduce post-harvest losses and improve profitability.

In this area, a great deal of data and expertise has been accumulated. This data is typically collected in the form of books and bulletins that span a period of time. Only through the presentations provided by scientists at meetings and conferences using power point presentations can up-to-date and succinct knowledge be obtained. In this book, we have attempted to compile our favorite presentations. We hope that this publication will be successful in spreading the message of food processing preservation and value addition, resulting in a rise in processing levels in the country and a reduction in post-harvest losses. For effective dissemination of this knowledge to end users, the book will be useful to scientists, academicians, students, policymakers, and particularly food technologists and extension staff.

Editors

CONTENTS

CHAPTER 1

FOOD PROCESSING PRINCIPLES AND APPLICATIONS

Anjum Ayoub*[1], Fozia Hameed[1], Jagmohan Singh[1], Kundan Veer Singh[2] & Nadira Anjum[1]

[1]*Division of Food Science and Technology, Sher-e-Kashmir University of Agricultural Sciences and Technoloy-Jammu, Chatha-180009, India*

[1]*Division of Plant Breeding and Genetics, Sher-e-Kashmir University of Agricultural Sciences and Technoloy-Jammu, Chatha-180009, India*

*Corresponding Author: *[1]anjumparay77@gmail.com@gmail.com*

ABSTRACT

The most critical feature of all food products is post-harvest processing and value addition. The term "processing" refers to the process of transforming raw food content into a consumer-acceptable type that has a longer shelf life and maintains nutritional quality. Drying, dehydration, evaporation, extrusion, freezing, and other traditional food processing methods may be used to achieve this. The loss of colour, flavour, vitamins, and other nutrients due to higher temperatures and longer processing times is the biggest flaw with these traditional methods. These methods also result in issues such as fouling on the heating surface, energy inefficiency, and the release of combustion gases into the atmosphere, resulting in emissions. Non-conventional food processing technologies such as ohmic heating, microwave heating, ultrasound, pulsed electric field, and high pressure processing are new processing and preservation methods that are emerging as green technologies in the food processing industry, resulting in energy consumption, water savings, and pollution reduction. These technologies can produce high-quality goods with improved heating performance, energy savings, and manufacturing time and costs. Pasteurization, sterilisation, oil extraction, juice extraction, clarification, filtration, frying, baking, blanching, evaporation, drying, and dehydration are only a few of the uses for these technologies in the food industry.

Keywords: Processing, Preservation, Non-conventional technologies, Microwave heating, Shelflife, quality.

INTRODUCTION

Processing of foods is a part of manufacturing industry that evolves animal, plant, and marine matter into intermediate or finished value-added food products that are safer to eat. This requires the use of labor, energy, machinery, and scientific knowledge to a step (unit operation) or a series of steps (process) in achieving the desired evolution (Heldman & Hartel, 1998). From the raw materials value-added finished products that satisfy consumer needs and convenience are obtained. According to Fellow, 2000, the aim of food treating is considered into four-fragments viz., expanding the period during which food remains wholesome (biochemical and microbial), value addition, convenience in diet and supplementing numerous kinds of nutrients for health demand. Storage life of foodstuff is raised by preserving the product against related to physical, chemical and biological. The three chief organisms of biological hazards are viruses, parasites and bacteria that may increase risk in treated foods. Presence of extraneous material such as stones, dirt, metal, glass, insect fragments, hair are physical hazards. chemical hazards that are present in food are pesticide, allergens, fungicide and mycotoxin. In the raw food part, biohazard mainly include pathogenic microbes. Such distinct hazards may accidentally become part of the treated product. Food treating actions warrant targeted removal of these hazards so that consumers enjoy safe, nutritious, wholesome foods. The scope of food treating is wide and advent of new techniques is ongoing still now and will in future. According to Heldman & Hartel (1998), general foodstuff processing actions contains handling of raw material, cooking, formulation constituent, heating and cooling, freezing, shaping and wrapping. These could widely have partitioned into primary and secondary processing. Primary processing occurs after harvest or slaughter and warrants that foods are easily transported and are ready to be sold, eaten or processed into other products (e.g. after the primary processing of peeling and slicing, an apple can be eaten fresh or baked into a pie). Secondary processing turns the primary-processed food or ingredient into other food products.

The traditional heat treating techniques, such as evaporation, pasteurization, sterilization, drying is still used in foodstuff sectors to treat the raw agricultural products. According to Vicente and Castro, (2007), traditional techniques relay on the generation of large amount of heat outside the product to be heated, by combustion of fuels or by an electric heated, which degrade the nutritional value of the product because of large extent of heat. Therefore, in order to maintain food quality, microbial safety, to raise working life of equipment and heat transfer efficiency and to reduce atmospheric pollution we have to adapt some modern food processing technologies that can efficiently replace the traditional heat foodstuff treating technique. Ohmic heating and microwave heating are emerging alternatives to traditional techniques of thermal treating which generate heat directly inside the food products without using any external heating surface. Novel techniques (non-thermal) have capability to deactivate enzymes and spores of microbesat reduced temperatures and hence preserves the nutritional and sensory quality of products (Stoica *et al.*, 2013).

General techniques for Preservation/ Treating of foodstuff

Goals of foodstuff treating

The foodstuff company uses a numerous kinds of techniques like freezing, refrigeration, dehydration, and thermal treating to preserve food materials. The Chief goal is to eliminate destructive pathogens present in the foodstuff and reducing or complete removal of spoilage to increase storage life. According to Heldman & Hartel, (1998), the usual fundamentals pertained with treating of foodstuff to raise storage life is to grab expansion and preserve quality by adding or removal of heat, removal of moisture and appropriate wrapping to maintain fruitful aspects for better treatment.

Foodstuff treating by addition or removal of heat

1. Pasteurization

Louis Pasteur's Thermal pasteurization is a relatively mild heat treatment, in which liquids, semi-liquids or liquids with particulates are heated at a specific temperature (often below 100° C) for a stated duration to destroy the most heat-resistant vegetativepathogenic organisms present in the food. This also results in shelf life extension of the treated product. According to Singh & Heldman (2009), Distinct temperature-time combinations can be utilizing to achieve pasteurization. pasteurization also applied for eradication of undesirable enzymes, such as lipase in milk. In contemporary years, the term "pasteurization" is also expanded to destroying pathogenic microorganisms in solid foods (such as pasteurization of almonds through oil roasting, dry roasting, and steam processing).As a moderate heat treatment, pasteurization usually affects sensory properties of foodstuff with limited storage life expansion, also pasteurized products require refrigeration as a secondary barrier for microbiological protection.

The intensity of thermal treatment required by a given product is affected by product pH; for example, fruit juices (pH <4.5) are generally pasteurized at 65° C for 30 min, compared to other low-acid vegetables that need to be treated at 121° C for 20–30 min.

2. Blanching

Blanching is generally applied as a pretreatment to thermal sterilization, dehydration, and freezing to control enzymes present in the food It is similar in temperature- time intensity to pasteurization, having a chief aim is to inactivate enzymes in fruits and vegetables that catalyze degradation reaction. It also destroys some microbes. depending on the kind of product used, blanching is done by applying boiling water or steam for short period of time (5-15 min. or more). Other boon effects are color enhancement and decrease discoloration. Other benefits of blanching include removal of air from food tissue and softening plant tissue to facilitate wrapping into food containers.

3. Thermal sterilization/ Canning or Retorting

According to Pflug (1998), Thermal sterilization involves heating the foodstuff to a satisfactorily high temperature (>100 ° C) and holding the product at this temperature for a specified duration, with the goal of inactivating bacterial spores of public health significance. Prolonged thermal exposure may degrade sensory and nutritional quality of the foodstuff.

4. Aseptic processing/ Microbial Free Environmental Foodstuff Treating

Traditional aseptic treating technologies use heat exchangers for example scrapped surface exchangers, but on the other hand, the enhanced foodstuff preservation technologies use ohmic or microwave heating instead of heat exchangers. Generally aseptic processing is a continuous thermal process, which involves pumping of food material through a set of heat exchangers where the product is rapidly heated under pressure to 130 ° C to make storage -stable foods. According to Sastry & Cornelius (2002), The heated product is then passed through a holding tube, wherein the temperature of the product mixture is equilibrated and held constant for a short period as determined by the type of foodstuff and microbes present, and passes through set of cooling heat exchangers to cool the product. The sterilized cooled product is then aseptically wrapped in a pre-sterilized wrapper.

5. Sous-vide cooking

It involves vacuum wrapping of foodstuff before use of low-temperature (65–95 ° C) heating and storing under refrigerated conditions (0–3 ° C).Due to use of modest temperatures, sous-vide cooking is not lethal enough to inactivate harmful bacterial spores. Example, fillet of salmon, fish stews, ready meals and meat. This technology is adapted and using in European countries.

6. Microwave heating

According to Ramaswamy & Tang (2008), Microwave energy generates heat in dielectric materials such as foods through dipole rotation and/or ionic polarization. In microwave heating, rapid volumetric heating could reduce the time required to achieve the desired temperature. For industrial application, a lower frequency of 915 MHz is selected for greater penetration depth.

7. Electro conductive heating (ECH)

It involves electrical resistance heating of the pumpable food to rapidly heat the food material. The heat is generated in the form of an internal energy transformation (from electric to thermal) within the material as a function of an applied electric field (<100 V/cm) and the electrical conductivity of the food (Sastry, 2008). application of electro conductive heating includes improvement of extraction, expression, drying, fermentation, blanching, and peeling.

- **Principle**- It is a simple food processing technique in which alternating electric current (AC) is passed through liquid or liquid-particulates foods primarily for heating. AC voltage is used to the electrodes at both ends of the product body.

- **Equipment**- The main components consist of electrodes, heating chamber and AC supply.It also contains lid, control panel, stand, and voltmeter and ampere meter.
- **Boon**-uniform heating, high energy efficiency, reduced risks of fouling of heat transfer surface and burning of the food, higher temperature in shorter time can be achieved, ease of process control with instant switch-on and shut-down and eco-friendly processing (Sakr and Liu, 2014).
- **Bane**-unsuitability for solid or dry foods and requirement voltage adjustments.

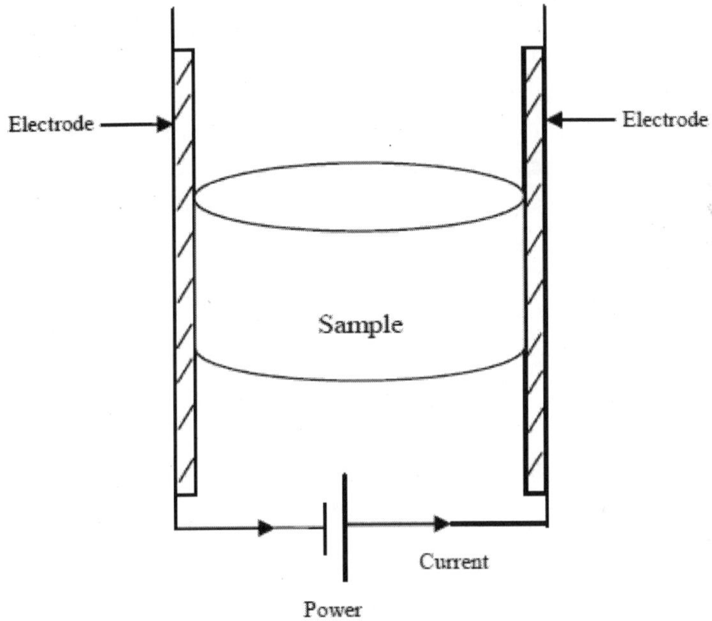

Fig 1. Electro Conductive Heating

8. *Drying*

According to Heldman & Hartel (1998), Drying is a process of moving the water present in the internal foodstuff matrix to its surface and then removing the surface water by evaporation.

It is one of the oldest methods of preserving foodstuff. It generally involves simultaneous heat and mass transfer. Most drying operations involve changing free water present within the food to vapor form and removing it by passing hot air over the product. During drying, the heat is transferred from an external heating medium into the food. The moisture within the food moves towards the surface of the material due to the vapor pressure gradient between the surface and interior of the product. The moisture is then evaporated into the heat transfer medium (usually air). The heat transfer can be accomplished through conduction, convection or radiation. There are various drying and dehydration methods frequently used in foodstuff treating such as hot air drying, spray drying, vacuum drying, freeze drying, osmotic dehydration, etc. During hot air drying, heat is transferred through the food either from heated air or from heated surfaces. Vacuum drying involves evaporation of water under vacuum or reduced pressures. Freeze drying involves removing the water vapor through a process called sublimation. Freeze drying helps to maintain food structure.

9. Refrigeration and freezing

Refrigeration and freezing have become a chief component of the food chain; depending on the type of product, they are used in all stages of the chain, from food processing, to distribution, retail, and final consumption at home. According to Tassou *et al.* (2010), The food industry employs both refrigeration and freezing processes where the food is cooled from ambient to temperatures above 0 ° C in the former and between 18 ° C and 35 ° C in the latter to slow the physical, microbiological, and chemical activities that cause deterioration in foods.

Non-thermal foodstuff treating and preservation

From the last two decades, due to raised purchaser interest in minimally processed foods with reduced preservatives, several non-thermal preservation methods have been investigated, these techniques generally use lethal agents (like U.V radiation, pulsed electric field, irradiation, pressure, ultrasound and others) with or without combination of heat to inactivate microbes. This assists to decrease the harshness of thermal exposure and preserve quality and nutrients. Since the mechanism of microbe inactivation by non-thermal lethal agents may be different from of heat, it is significant to understand the synergy, additive, or antagonistic effects of sequential combinations of distinct lethal agents.

1. Irradiation

It is one of the utmost widely studied non-thermal technologies. Ionizing radiation (I.R) consists of γ-ray and electron beam. During irradiation of foodstuffs, I.R penetrates a food and energy is absorbed. Absorbed dose of radiation is expressed in grays (Gy), (where 1 Gy is equal to an absorbed energy of 1 J/kg). Milder doses (0.1–3 kGy), called "radurization," are used for shelf life extension, control of ripening, and inhibition of sprouting.For storage life expansion, to govern ripening and inhibition sprouting, in foodstuff sector, milder doses (0.1-3 kGy) are applied which is commonly known as radurization. To reduce viable non-spore forming pathogenic bacteria, Radicidation often used between a dose of 3 to 10 kGy while for sterilization of bacterial spores, Radappertization is applied with a dose of 10kGy to 50 kGy. From its beginning in the 1960s, the symbol Radura has been used to indicate ionizing radiation treatment (Figure 2).

According to Molins, (2001), Radiation is quite operative in penetrating through various wrapping materials. However, the radiation dose may cause changes in packaging polymers. Thus, careful choice of wrapping material is critical to avoid any radiolytic products from packaging contaminating the food products.

Fig.2- International Radura symbol for irradiation on the packaging of irradiated foods

2. Ultra High-pressure processing

To achieve microbial inactivation foodstuff, high hydrostatic pressure processing or simply high pressure processing, uses elevated pressures (up to 600 MPa), with or without the addition of external heat (up to 120 ° C).Pressure pasteurization treatment (400–600 MPa at chilled or ambient conditions), usually, has bane the effects on color and nutrition. Uniform compression heating and expansion cooling on decompression assists to decrease the harshness of thermal effects such as quality degradation and nutritional loss encountered with traditional treating technologies. Examples-smoothies, poultry products, oysters, ham, fruit juices, and salsa.

3. Pulsed electric field (PEF) processing

- **Principle**: Typical PEF equipment components include pulse generators, treatment chambers, and fluid-handling systems, as well as monitoring and control devices. During PEF processing, a high voltage electrical field (20–80 kV/cm) is applied across the food for a few microseconds. A number of process parameters including electric field strength, treatment temperature, flow rate or treatment time, pulse shape, pulse width, frequency, and pulse polarity govern the microbiological safety of the processed foods.
- **Equipment:** Typical PEF system consists of a high-voltage power supply, capacitors to store the charge, a discharge switch to release the charge to electrodes and a pump to force the liquid product from feed tank to treatment chamber.

- **Uses :** It has many applications in food industry for liquid foods which includes pasteurisation and extraction of fruit juices (Korma *et al.*, 2016),soups, liquid egg (Nieto *et al.*, 2003) and milk (Hawa *et al.*, 2011), decontamination of heat sensitive foods (Zhao *et al.*, 2008).
- PEF technology has the potential to pasteurize a variety of liquid foods including fruit juices, soups, milk, and other beverages.
- **Boon-**widely used to kill vegetative cells, to preserve colour, flavour and nutrients of the product without any toxicity and to process the food in relatively short treatment time.
- **Bane-**difficult to use with conductive materials, no effect on enzymes and spores, suitable for liquids or particles in liquids only and products of electrolysis may adversely affect foods.

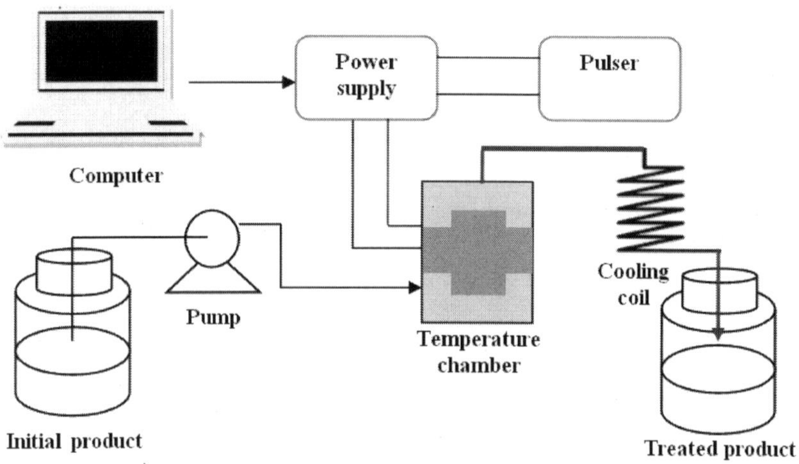

Fig 3. Pulse electric field (PEF)

4. Ultrasound

Ultrasound is a form of energy generated by sound waves of frequencies above 16 kHz; when these waves propagate through a medium, compressions and depressions of the medium particles create microbubbles, which collapse (cavitation) and result in extreme shear forces that disintegrate biological materials. According to Piyasena *et al.*, (2003), High-power ultrasound processing or sonication is another alternative technology that has shown promise in the food industry, especially for liquid foods, in inactivating spoilage microorganisms.

- **Principle-** Sound is vibration that transmitted in a medium and can hear by human ear. Sound waves having frequencies greater than 20 kHz are called "Ultrasounds". Ultrasounds are classified into two types, low energy ultrasounds and high energy ultrasound. Low energy ultrasounds have the frequencies greater than 100 kHz, low power and low-intensity (<1 W/cm).

These are non-damaging, and uses slight power that cannot cause physical or chemical manipulations in the materials property and can used for providing information about physicochemical properties of foods like composition, structure and physical state of the food.

High energy ultrasound on the other hand have the frequencies 18-100 kHz, high power and high-intensity (>1 W/cm) (Villamiel and de Jong, 2000).

These waves cause physical, mechanical or chemical effects on material and can be used for physical disruption and acceleration of chemical reactions (Awad *et al.*, 2012).

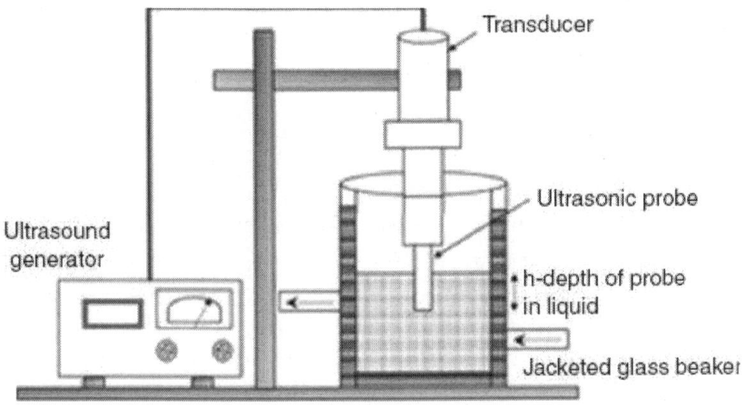

Fig 4. Ultra-Sound

- **Equipment :** The basic components of ultrasonic systems are power generator, transducer and reactor. Power generator is nothing but the source of electrical power supply. The transducers are the devices used to convert electrical energy, coming from a power generator, into mechanical energy in the form of ultrasonic vibrations. The ultrasonic bath and probe are reactors.
- **Uses :** In Food sector- chemical, mechanical and biochemical effects like crystallization (Chow *et al.*, 2003), defoaming, drying, freezing and tenderization of meat (Knorr *et al.*, 2004), extraction (Li *et al.*, 2004), degassing, filtration (Muthukumaran *et al.*, 2006), fermentation (Matsuura *et al.*, 1994), mixing, emulsification/homogenization (Freitas *et al.*, 2006), effluent treatment, enzyme and microbial inactivation, sterilization of equipments etc (Chemat *et al.*, 2011).
- **Boons :** Effectiveness against vegetative cells, spores and enzymes, reduced processing times and temperatures, little adaptation required of existing processing plant, increased heat transfer and can be both batch or continuous operation.
- **Banes :** Depth of penetration is affected by solids and air in product, possible damage by free radicals, unwanted modification of food structure and texture

may possible, sometimes need to be used in combination with another process like heat or pressure and potential problems with scaling-up plant.

Other foodstuff preservation/ treating techniques

1. Fermentation

In terms of nutrition or digestion, Fermentation causes desirable biochemical alterations in foodstuffs few examples include cheese, yogurt, alcoholic beverages, salami, beer, and pickles etc. Representative vegetative bacteria in the fermentations are Lactobacillus, Lactococcus, Bacillus, Streptococcus, and Pseudomonas spp. Yeast and fungi (e.g. Saccharomyces, Endomycopsis, and Monascus) are also used for fermentation.

2. Extrusion

According to Singh & Heldman, (2009), Extrusion is a process that converts raw material into a Product with a desired shape and form, such as pasta, snacks, textured vegetable protein, and ready-to-eat cereals, by forcing the material through a small opening using pressure. According to Riaz, (2000), Some of the specific boon include high productivity, adaptability, process scale-up, energy efficiency, low cost, and zero effluents.

3. Baking

It applies dry heat to cook fully enhanced flour dough into a distinct kind of baked products such as bread, cake, pastries, pies, cookies, scones, crackers, and pretzels. The dough needs to undergo various stages (mixing, fermentation, punching/ sheeting, panning, proofing among others) before it is ready for baking (Figoni, 2010).

4. Hurdle Technique

Hurdle technology involves a suitable combination of different lethal agents to ensure microbial safety, quality, and stability of the processed product. Heat, pressure, acidity, water activity, chemical/natural preservatives, and packaging are examples of hurdles that can be combined to improve the quality of the final processed product. The hurdle approach requires the intensity of individual lethal agents (for example, heat or pressure) to be relatively modest, yet is quite effective in controlling microbial risk.

5. Wrapping

It plays a pivotal role in many food preservation operations, has numerous roles, including containment, preservation, communication/education, handling/ transportation, and marketing. Wrapping assists in maintain during storage the quality and properties of foods attained via processing. It protects the food material from microbiological contaminants and other environmental factors. The package also helps prevents light-induced changes in stored food products and minimizes loss of moisture.

REFERENCES

- Fellow, P. 2000. Food processing technology principles and practice. Second eds. In-Processing using electric fields, high hydrostatic pressure, light or ultrasound, Woodhead Publishing Limited and CRC Press LLC. Pp. 210-227.
- Stoica, M., L. Mihalcea, D. Borda and Alexe, P. 2013. Non-thermal novel food processing technologies. An overview. *Journal of Agroalimentary Processes and Technologies*. **19**(2)**:** 212-217.
- Vicente, A. and Castro, I. 2007. Novel thermal processing technologies. In G. Tewaria and V. Juneja (Eds.), Advances in thermal and non-thermal food preservation. Oxford, UK: Blackwell Publishing.
- Heldman DR, Hartel RW (1998) Principles of Food Processing. New York: Springer.
- Singh RP, Heldman DR (2009) Introduction to Food Engineering, 4th edn. San Diego, CA: Academic Press.
- Pflug IJ (1998) Microbiology and Engineering of Sterilization Processes, 9th edn. Minneapolis, MN: Environmental Sterilization Laboratory.
- Sastry SK, Cornelius BD (2002) Aseptic Processing of Foods Containing Solid Particulates. New York: John Wiley.
- Ramaswamy HS, Tang J (2008) Microwave and radio frequency heating. *Food Science and Technology International***14:** 423–428.
- Sastry SK (2008) Ohmic heating and moderate electric field processing. *Food Science and Technology International***14:** 419–422.
- Sakr, M. and Liu, S. 2014. A comprehensive review on applications of ohmic heating (OH). Renewable and Sustainable Energy Reviews. **39:** 262–269.
- Tassou SA, Lewis JS, Ge YT, Hadawey A, Chaer I (2010) A review of emerging technologies for food refrigeration applications. Applied Thermal Engineering **30:** 263–276.
- Molins RA (ed) (2001) Food Irradiation: Principles and Applications. New York: John Wiley.
- Korma, S. A., K. Alahmad, A. H. Ali, M. Shoaib, S. M. Abed, H. Yves, J. N. Atindana and Qin, J. 2016. Application of pulsed electric field technology in apple juice processing. *Austin J Nutri Food Sci.***4**(2)**:** 1080.
- Nieto, G. M. M., P. D. Pedrow, B. G. Swanson and Barbosa-Canovas G. V. 2003. Energy analysis of liquid whole egg pasteurized by pulsed electric fields. *Journal of Food Engineering*. **57**(3)**:**209–216.

- Hawa, L. C. R. I. Putri and Susilo, B. 2011. Pulsed Electric fields pasteurization of milk: Effects of various voltage and treatment time on physical properties. *J. Basic. Appl. Sci. Res.* **1**(10)**:**1516-1523.
- Zhao, W., R. Yang, R. Lu, M. Wang, P. Qian and W. Yang. 2008. Effect of PEF on microbial inactivation and physical–chemical properties of green tea extracts". *Food Science and Technology.* **41**(3)**:** 425-431.
- Piyasena P, Mohareb E, McKellar RC (2003) Inactivation of microbes using ultrasound: a review. *International Journal of Food Microbiology***87:** 207–216.
- Villamiel, M. and de Jong, P. 2000. Inactivation of Pseudomonas fluorescensand Streptococcus thermophilus in trypticase soy broth and total bacteria in milk by continuous-flow ultrasonic treatment and conventional heating. *Food Eng.*, **45**(3)**:** 171– 179.
- Awad, T. S., H. A. Moharram, O. E. Shaltout, D. Asker and Youssef, M. M. 2012. Applications of ultrasound in analysis, processing and quality control of food: A review. *Food Research International.* **48**(2)**:** 410–427.
- Chow, R., R. Blindt, R. Chivers and Povey, M. 2003. The sonocrystallization of ice in sucrose solutions: primary and secondary nucleation. Ultrasonics, **41**(8)**:** 595 604.
- Knorr, D., M. Zenker, V. Heinz and Lee, D. U. 2004. Applications and potential of ultrasonics in food processing. *Trends in Food Science and Technology.* **15**(5)**:** 261–266.
- Li, H., L. Pordesimo and Weiss, J. 2004. High intensity ultrasound-assisted extraction of oil from soybeans. *Food Research International*. **37**(7)**:** 731–738.
- Muthukumaran, S., S. E. Kentish, G. W. Stevens, and Ashokkumar, M. 2006. Application of ultrasound in membrane separation processes: A review. *Reviews in Chemical Engineering*. **22**(3)**:** 155 194.
- Matsuura, K., M. Hirotsune, Y. Nunokawa, M. Satoh and Honda, K. 1994. Acceleration of cell growth and ester formation by ultrasonic wave irradiation. *Journal of Fermentation and Bioengineering*.**77**(1)**:**36 40.
- Freitas, S., G. Hielscher, H. P. Merkle and Gander, B. 2006. Continuous contact and contamination free ultrasonic emulsification- A useful tool for pharmacuetical development and production. Ultrasonics Sonochemistry. **13**(1)**:**76 85.
- Chemat, F., Zill-e-Huma and Khan, M. K. 2011. Applications of ultrasound in food technology: processing, preservation and extraction. Ultrasonics Sonochemistry. **18**(4)**:** 813–835.
- Riaz MN (2000) Extruders in Food Applications. Boca Raton, FL: CRC Press.
- Figoni P (2010) How Baking Works: Exploring the Fundamentals of Baking Science, 2nd edn. New Jersey: Wiley.

2

CHAPTER

CONVENTIONAL TECHNOLOGIES FOR FOOD PRESERVATION

Fozia Hameed[*1], Anjum Ayoub[1], Neeraj Gupta[1] & Nadira Anjum[1]

*[*1]Division of Food Science and Technology, Sher-e-Kashmir University of Agricultural Sciences and Technoloy-Jammu, Chatha-180009, India*

*Corresponding Author: [*1]s.fozia011@gmail.com*

ABSTRACT

Food preservation has been an essential activity throughout the human history. It aims at preventing and reducing the loss of food in the production system and extending its shelf life. A good method of food preservation is one that slows down or prevents altogether the action of the spoilage agents without damaging the food. To achieve this, different preservation technologies such as drying, curing, pickling, fermentation, sugaring, pasteurization, freezing, blanching and refrigeration have been fostered with time. Preservation also makes it possible to consume some of the foods during off seasons, throughout the year. These conventional technologies not only increase the shelf-life of the foods but also minimize the risk, protect the environment and minimize the quality degradation by microbial, enzymatic and chemical reactions.

Key words: food preservation, conventional technologies, shelf-life, consumer

INTRODUCTION

Food is an organic edible substance that can be raw, processed or formulated in solid, semi-solid or liquid form consumed orally for growth, nutrition, pleasure and satisfaction (Rahman, 2007). Foods are mainly composed of edible bio-chemicals such as water, carbohydrates, proteins, lipids, minerals and other organic substances. These organic substances include vitamins, acids, pigments, flavoring compounds, emulsifiers, antioxidants and polyphenols (Prokopov and Tanchev, 2007). The composition of food basically determines what actually we eat in a food and thus acts as a critical bridge between nutrition, health promotion, disease prevention and food production (Pehrsson and Haytowitz, 2016). Foods can be classified as fresh and minimally processed, perishable and non-perishable, preserved, manufactured and formulated, primary and secondary derivative and synthetic, functional and medicinal foods. All kinds of foods undergo spoilage or deterioration due to a wide range of reactions including physical, chemical, enzymatic and microbial (Gould, 2000). Therefore, various types of spoilage can be prevented to a greater extent by numerous preservation techniques, mostly by slowing or inhibiting the microbial growth and enzymatic activity. Since time immemorial, civilized men adopted certain techniques to preserve the food. These include curing, freezing, refrigeration, salting, syruping, canning, pickling and fermentation (Kumar et al., 2019). Thus, the processes or techniques undertaken in order to control both intrinsic as well as extrinsic factors that may cause spoilage without damaging the quality of food is called food preservation. Preservation of foods not only increases the shelf-life but also retains the nutritional value as well as color, texture and flavor of the food commodity (Amit et al., 2017). Further, the preservation method mostly depends on the type of food that needs to be processed or formulated.

Food preservation dates back to prehistoric ages when the ancient man felt the necessity of food storage and preservation especially after slaughtering a big animal which was not possible to eat at a time. Knowing the techniques or processes of preserving food such as by freezing the sealed meat on the ice or by drying the foods in the sun was the first and foremost step towards civilization (Nummer, 2002). The primitive troupe no longer had to consume the hunt or harvest immediately, instead could preserve some for later use. Each culture at different times and locations used almost the similar basic techniques for food preservation. Further, the cycle of season brings periods of shortage and abundance of different foods at different times of the year. Therefore, preservation makes it possible to consume such foods during off-seasons. Understanding the effects of each preservation method has also become essential in all aspects (Prokopov and Tanchev, 2007). This chapter presents the basic principle of food preservation, discusses various internal and external factors responsible for spoilage of food, provides an overview of the conventional technologies used for preserving food items and also identifies the changes in food quality.

2. OBJECTIVES OF FOOD PRESERVATION

Foods are perishable in nature and deteriorate more often, becoming unsafe or unacceptable for human consumption. So, the primary objective of the food preservation is to prevent food spoilage until it can be consumed. Preserving foods also offers the opportunity to have a wide variety of foods year around. Further, food preservation is economical because it includes saving money as well as satisfying personal preferences. Thus, food preservation is mainly done to i) extend the shelf-life of food ii) preserve the safety and acceptability iii) retain the nutritive value as well as economic viability (Srivastava and kumar, 2003).

2. PRINCIPLES OF FOOD PRESERVATION

Food preservation basically consists of the application of science-based knowledge to prevent food and food product spoilage and extend their shelf-life, also ensuring the consumers at the same time that the product retains its nutritional value and is free of pathogenic microorganisms (Fellows, 2009). Control of spoilage by any means becomes necessary to increase the shelf life (Amit et al., 2017). Spoilage starts the time animals are slaughtered and plant foods are harvested. This involves two main processes 1) autolysis 2) microbial growth. What actually happens is cells start breaking down due to the presence of enzymes in food i.e, self-destruction of cells occurs. These disintegrated cells are exposed to the microbial activity and later causes changes in color, odor, texture and flavor of food. For effective food preservation, it thus becomes necessary to prevent both cell-destruction as well as microbial growth. Therefore, the basic principle of food preservation includes:

3.1 PRESERVATION OR DELAY OF MICROBIAL DECOMPOSITION

Several preservation techniques prevent microbial growth and decomposition of foods. This can be achieved by lengthening the lag phase and generation time. When all microbes are removed or killed, decomposition is prevented. Microbial growth and decomposition can be prevented or delayed by:

3.1.1 Keeping out microorganisms (Asepsis)

Asepsis is the process of keeping out microorganisms as a preservative factor. This includes either avoiding addition of spoilage microbes from unclean processing equipments, containers or utensils or restricting introduction of actively growing microbes to the max. Packaging of foods especially heated fruit juices by aseptic techniques is a widely used application of asepsis (Srivastava and Kumar, 2003).

3.1.2 Removal of microorganisms

Removal of microorganisms is not very effective in food preservation but under special conditions it may be helpful. Microorganisms can be removed from the foods by either of the following techniques (Srivastava and Kumar, 2003):

3.1.2.1 Filtration

It is the only successful method for the removal of microorganisms. Its use is limited to clear liquids such as fruit juices, beer, soft drinks, wine and water.

3.1.2.2 Centrifugation

This method is not very effective as all microorganisms are not removed. Centrifugation is used in the treatment of drinking water but is insufficient by itself.

In case of milk, the main reason behind centrifugation is to take out suspended materials not to remove bacteria. However, bactofugation is used to remove bacteria from milk.

3.1.2.3 Washing

Washing raw foods can remove a number of spoilage microorganisms from the surface. Washing fresh fruits and vegetables reduces soil microorganisms on the surface. On the other hand, washing of foods may prove dangerous in case water adds microorganisms or increase the moisture.

3.1.2.4 Trimming

Trimming away spoiled part of a food or discarding spoiled samples of food may reduce microorganisms from food.

3.1.3 Hindering the growth and activity of microorganisms

This can be archived by low temperature, pH, drying additives and anaerobic condition. Unfavorable condition extends lag phase and generation time, thus, will increase the shelf life of food (Srivastava and Kumar, 2003).

3.1.4 Killing microorganisms

Microbial load can be reduced by killing or sub-lethally injuring the microorganisms. This can be achieved by heat or radiation. Sub-lethally injured cells require better culture medium for growth than the untreated or non-injured cells. Microbial inactivation in foods will increase the shelf life of the food.

3.2 PRESERVATION OR DELAY OF SELF-DECOMPOSITION OF FOOD

This is accomplished by i) destruction or inactivation of food enzymes i.e, by blanching ii) delay of chemical reactions 1.e, preservation of oxidation by means of an antioxidant (Srivastava and Kumar, 2003).

3.3 PREVENTION OF DAMAGE CAUSED BY INSECTS, ANIMALS, AND MECHANICAL CAUSES

The insects, animals and other organisms can damage foods during transportation and storage. Processing of foods such as freezing, drying, pressure can cause mechanical damage to foods (Srivastava and Kumar, 2003).

4. FACTORS THAT CAUSE THE SPOILAGE OF FOOD

Food spoilage can be defined as a process which renders the food product unacceptable for consumption. This phenomenon is the outcome of the biochemical activity of microbial chemical processes. Spoilage can occur at any stage along food chain i.e, production, transportation, processing, distribution and storage (Amit et al., 2017). Food spoilage is linked to food safety. The primitive stage of food spoilage can be detected by color, smell, flavor and texture of food product. Spoiled foods may be safe to eat i.e, they may not cause illness because of no pathogens and toxins present. However, changes in organoleptic properties cause them to be rejected. Spoilage of foods may arise from physical damage, chemical reactions, indigenous enzymatic activity and microbial actions (Amit et al., 2017). Different factors that affect the food spoilage are briefly discussed in the following sections.

4.1 PHYSICAL SPOILAGE

Physical food spoilage results when moist foods are excessively dehydrated or dried foods absorb excessive moisture. Thus, food spoilage due to physical changes or instability is defined as physical spoilage. Examples of physical spoilage include moisture loss or gain, moisture migration between different components and physical separation of components (Amit et al., 2017). The factors affecting physical spoilage are moisture content, temperature, glass transient temperature, crystal growth and crystallization.

4.1.1 Moisture content

Change in the water content i.e, loss, gain or migration is the major cause of food spoilage (Fabunmi et al., 2015). It is directly related to the water activity (aw) of food product (Steele, 2004). Water activity is a thermodynamic property which is defined as the ratio of vapor pressure of water in a food substrate to the vapor pressure of pure water at the same temperature (Barnwa et al., 2010). It determines the shelf stability and safety of food products. Water activity of pure water is exactly 1. As temperature increases, aw typically increases, except in products with crystalline salt and sugar. Water activity is used for the preservation of food, stabilization of the food supply and developing different types of the shelf-stable foods. Lower the water activity, longer the shelf life. Water activity values can also help limit moisture migration within a food product made with different ingredients.

4.1.2 Temperature

The most important factor in case of fruit and vegetable spoilage is temperature. An optimum temperature range is given for slow ripening and to maximize post-harvest life. Slow ripening also needs an optimal relative humidity and optimum air movement around fruit and vegetable. Apparently, these optimum conditions are called modified atmospheres (MA). Temperature usually besets the metabolism of the commodities and contemporarily alters the rate of attaining desired MA (Kader et al., 1989). Low temperature can also have a negative effect on foods that are susceptible to freeze damage. At a lower temperature, when food products become partially frozen, breakage in cells occurs which damages the product. Most tropical fruits and vegetables are sensitive to chilling injury. This generally occurs before the food product starts to freeze at a temperature in between 5 C and 15 C (Steele, 2004).

4.1.3 Glass transition temperature

Another factor that affects the shelf life of food and food products is glass transition temperature. A food is most stable at and below its glass transition temperature. Glass transition mainly depends on the food composition, temperature, relative humidity, concentration of water and other plasticizers (White and Cakebread, 1966). Physical stability of foods is related to glass transition temperature. When dry food products are kept in high humid conditions, change in state of the food product occurs as a result of glass transition phenomenon (Steele, 2004). Low glass transition means food is soft and relatively plastic while as high glass transition temperature means food may even flow.

4.1.4 Crystal growth and crystallization

Crystals often play an important role in food product quality and shelf life. Foods undergoing slow freezing suffer severely due to large crystal growth. Controlling crystallization to obtain the desired crystal content, size distribution, shape and polymorph is key to manufacturing products with desired functionality and shelf life. To minimize the large crystal growth, emulsifiers and other water binding agents can be added during freezing cycles (Levine and Slade, 1988).

Foods with high sugar content can undergo sugar crystallization either by moisture accumulation or by increasing temperature. As a consequence, sugar comes to the surface from inside, and a gray or white appearance is noticed. Staling of sugar cookies, graininess in candies, and ice creams are the results of sugar crystallization (Steele, 2004). Sugar crystallization can be delayed by the addition of fructose or starch. Moreover, above the respective glass-transition temperature, time plays a crucial role in sugar crystallization process of food items (Roos and Karel, 1991).

4.2 CHEMICAL SPOILAGE

Chemical food spoilage occurs when different components in the food react with each other or with some added component which alter the food's sensory characteristics (Petruzzi et al., 2017). Examples of this include: oxidation; enzymatic browning; and non-enzymatic browning. Except oxidation which is purely chemical based phenomena, chemical food spoilage is somehow interrelated with microbial actions such as putrefaction, pectin hydrolysis etc (Amit et al., 2017).

4.2.1 Oxidation

Amino acids occur naturally in certain foods such as dairy products, meats, fish, poultry and so on. These amino-acids get converted into organic acid and ammonia in presence of oxygen. This is the elementary spoilage reaction in case of refrigerated fresh meat and fish (Jay, 2000).

Amino acid + $O_2 \rightarrow NH_3$ + Organic acid

Fats are the esters of glycerol and fatty acids and are thus called glycerides. Fats undergo oxidation, hydrolysis, chemical changes that produce both desirable and undesirable flavor changes. The term rancidification is used to determine lipid oxidation in which polyunsaturated fats undergo reaction with oxygen resulting in unpleasant taste and odor, color alteration and toxic substances formation. The end-product of racidification is the formation of carbonyl compounds, responsible for rancid taste of foods. Rancidification can be catalyzed by the presence of metal oxides and exposure to light increases the reaction rate (Enfors, 2008).

4.2.2 Non-enzymatic browning

Another cause of food spoilage is non-enzymatic browning, also called as Maillard reaction. This reaction occurs between reducing sugars and proteins or free amines (Amit et al., 2017). Non-enzymatic browning occurs during the processing and storage of dehydrated and semi-moist foods such as dried milk, dried whole eggs and breakfast cereals, leading to color darkening, protein insolubility, bitter flavor and subsequent nutritional loss (Desrosier and Singh, 2014).

4.2.3 Enzymatic browning

Enzymatic browning is another cause of quality loss in fruits and vegetables. This reaction requires the action of enzymes and oxidation in order to occur. Oxygen in the air can cause fruits to brown specially when sliced or squashed. Fresh fruits and vegetables keep enzymes trapped in their tissues. However, when the fruit or vegetable is sliced, the enzyme come in contact with oxygen in the air and thus causes the fruit to turn brown.

Many important deteriorative reactions in food also are caused by enzymes. Enzyme-catalyzed processes may contribute quality deterioration, particularly in processed foods. For example, for orange juice, a stable colloidal suspension is desired, the action of pectin methyl esterase (PME) is undesirable since demethylation of pectin, catalyzed by this enzyme, leads to separation of serum. Hydrolases (cause hydrolytic reaction) and oxidoreductases (cause oxidation reduction) are the important enzymes causing food spoilage. Hydrolytic enzymes break chemical bonds that involve water in the mechanism (Erkmen and Bozoglu, 2016). Alfa-amylases hydrolyze -1-4 glucosidic bonds within starch resulting in dextrins, maltose, and maltotriose. Beta-amylase hydrolyzes -1-6 glycosidic bonds from the non-reducing end of starch and results in maltose. Glucoamylase hydrolyzes -1-4 and -1-6 glucosidic bonds in starch to produce glucose. These enzymes can be naturally present in foods or are produced by microorganisms (Erkmen and Bozoglu, 2016). Glucose oxidase catalyzes glucose oxidation by consuming oxygen from the air to D-gluconic acid. Lipoxygenases catalyze the oxidation of lipids in the presence of oxygen. It initiates the oxidative rancidity of plant lipids containing high proportion of unsaturated fatty acids and causing off-flavors (Erkmen and Bozoglu, 2016). Examples of enzymes and their contribution to quality loss in specific foods are summarized in Table 1.

Table 1. List of enzymes causing food spoilage

Enzymes	Type of spoilage
Alkaline protease	Heat-stable alkaline proteases can cause gelation in milk products processed at UHT
Ascorbic acid oxidase	Ascorbic acids are oxidized, and cause loss of vitamin C fruits, vegetables and juices
Cellulase	Results in loss of texture integrity and softening of plant foods, and releases glucose
Chlorophyllase	Removal of the phytol side chain from chlorophyll causes the degreening on planty foods (such as green vegetables)
Lipase	Hydrolyzes milk fat to release short-chain fatty acids and cause rancidity and off-flavors (such as in milk and meat)
Lipoxygenase	Formation of hydroperoxides can cause bleaching of pigments, offensive flavor causes texture change, destruction of vitamin A, and loss of nutrients (such as on legume seeds)
Pectinases	Destruction of pectic substance and this results in loss of texture integrity and softening (such as in vegetables and fruits).
Peroxidase	Decomposition o fhydrogen peroxides with generation of free radicals causes browning, off-flavor, and so on (such as in vegetables and fruits).
Phospholipase	Denaturation of muscle proteins and texture deterioration (such as in fish)
Proteinases	Decomposition of proteins with formation of amino acid and small peptides to give bitter flavor, reduction of shelf life, over tenderness (such as in eggs, fish, and flour)
Polyphenoloxidase	Browning, off-flavor, and vitamin loss (such as in fruits, vegetables, and shellfish)

4.3 MICROBIAL SPOILAGE

Microbiological food spoilage is caused by the growth of microorganisms which produce enzymes that lead to objectionable by-products in the food. The main microorganisms responsible for spoilage include yeast, mold and bacteria (Petruzzi et al., 2017). Storage of raw milk for long time at low temperature before heat treatment causes the psychrotrophic bacteria to multiply and produce extracellular heat-stable enzymes such as proteinases and lipases. These enzymes cause noticeable hydrolysis of proteins and lipids leading to the spoilage of milk within 3-7 days of storage. The bacterial enzymes also cause off-flavors and texture defects in raw fish and meat. However, factors that affect the microbial spoilage in foods are generally divided into two categories: i) intrinsic factors such as food structure, water activity (a_w), food composition and oxidation-reduction potential ii) extrinsic factors such as relative humidity, temperature, presence and activities of other microbes (Jay, 2000; Steele, 2004; Doyle, 2009).

4.3.1 Food structure

Raw foods have a structural integrity such as peel, shells, bran, collagen, tissue etc. that protects most of their mass from microbial contamination. But as soon as processing begins, the food becomes more vulnerable to the attack of microorganisms (Baron and Gautier, 2016). Liquid and semi-solid foods spoil rapidly by means of their own motility while as solid foods tend to spoil from outside towards inwards (Modi, 2009).

4.3.2 Food composition

The nutritional composition of the food determines the type of microbial spoilage. Foods rich in proteins such as meat, fish and poultry are usually attacked by proteolytic organisms. Carbohydrate foods are liable to spoilage by fermentative microbes. Similarly foods rich in fat are attacked by lipolytic organisms (Modi, 2009).

4.3.3 Water activity

Water activity of foods plays an important role for microbial growth as aqueous phase is the main element for their metabolic activity. Bacteria require high water activity and therefore cannot spoil dried foods except osmoprotectants that keep water inside the cytoplasm (Baron and Gautier, 2016). Yeasts have water activity above 0.87-0.94, but some osmotolerant species grow at water activity level 0.60. Molds can grow in foods with the lowest water activity i.e, 0.70-0.80; *Xeromyces* is one species of mold that can grow in water activity as low as 0.60 (Davidson and Critzer, 2012).

4.3.7 Temperature

Another important factor that influences the microbial spoilage is temperature. A higher temperature condition reduces the period of lag phase and generation times, leading to an increased growth rate (Kreyenschmidt and Ibald, 2012). Molds and yeasts tend to grow best at room temperature and below, and therefore assume greater importance in the spoilage of foods held at cool and chilled temperatures (Modi, 2009). Bacterial spores are among the most resistant, as well as certain fungal spores that can survive treatments lasting several hours at temperatures above 100 °C (Baron and Gautier, 2016). On another level, most microorganisms are more resistant to freezing when the process is rapidly carried out. This is because of the formation of intercellular microcrystals that do not alter the cells as much as large crystals formed during a slow freezing process. Whatever the freezing method, cell structures suffer deformation, damage, and destruction, which causes cell death at a rate of 1 Log. In general, Gram-positive bacteria are more resistant than Gram negative bacteria (Baron and Gautier, 2016).

5. CONVENTIONAL METHODS OF FOOD PRESERVATION

As already discussed, food preservation is a process or technique employed to prevent food spoilage or food borne illness and to enhance the shelf-life of foods. The following section gives a brief description of conventional food preservation technologies currently used by the food processing industries.

5.1 PRESERVATION BY DRYING

Drying is one of the oldest techniques of food preservation. It can be defined as the process in which water can be removed from the food by means of vaporization, thus reducing the water available for degradation reactions of chemical, enzymatic and microbial nature (Guine, 2018). Drying basically is a combination of heat and mass transfer for which energy must be supplied. It involves the use of heat to vaporize the water present inside the food as well as removal of water vapors from the food surface. The main objective of drying is to decrease the water activity (a_w) of various perishable materials to values <0.5, in order to enable their storage at ambient temperature (Amit et al., 2017). The aw is important in controlling the shelf-life of foods by suppressing the growth of micro-organisms, by reducing the rates of chemical reactions, and by inhibiting enzymatic deterioration.

Water activity determines the stability of a food than the total amount of water present. For most foods, the critical point below which no micro-organism can grow is in the 0.6–0.7 water activity range. A food product is most stable at its monolayer moisture content, which varies with the chemical composition and structure. Water activity is also useful for predicting the final moisture content at equilibrium versus drying conditions (temperature and relative humidity), and for the selection of ingredients and packaging method (Bonazzi and Dumoulin, 2011).

Advantages of drying

- Reduces weight and volume of foods
- Facilitates food storage, packaging and transportation
- Provides different flavors and aroma
- Increase the shelf-life of foods by reducing the moisture content

Despite all these benefits, drying also has certain limitations like significant loss of flavor, aroma and certain functional compounds like vitamin C, vitamin B1, proteins and lipids (Agrahar-Murugkar and Jha, 2010; Jangam et al., 2010).

5.2 PRESERVATION BY CURING

Curing can be defined as food preservation and flavoring process in which combinations of salt, sugar, nitrate and nitrite is added to the variety of foods such as meat, fish, vegetables etc. so as to draw moisture out of the food by means of osmosis (Kumar, 2019). The main purpose of curing is to increase the solute concentration and decrease the water potential in a food, thus making the food inhospitable for the microbial growth which causes food spoilage. Nitrate and nitrite in combination with salt inhibit the growth of *C. botulinum* and also contribute to the characteristic pink color of cured meats. Salt is also capable of slowing oxidation process resulting in slow oxidation of fat thus avoiding rancidity (Kumar, 2019).

5.3 PRESERVATION BY PICKLING

Another method of food preservation is pickling. Pickling is the process of preserving edible products in an acid solution usually vinegar or vegetable oil or in salt solution known as brine (Montano et al., 2016). Pickling can be divided into two categories: chemical pickling and fermentation pickling. In chemical pickling the food is placed in an edible solution like brine, vinegar, alcohol and vegetable oil that inhibits or kills bacteria or other organisms (Ockerman and Basu, 2014). On the other hand, fermentation pickling involves microorganisms such as starter culture or natural culture that on growth lower the pH to such a point that bacteria will not grow (Lee, 2004). In case of brining, the acid responsible for the preservative action is mainly lactic acid produced by fermentation process. The process of pickling is also known as brining and the resulting foods as pickles. Usually pickles refer to the vegetable products but sometimes fish, eggs or meat is also subjected to pickling (Montano et al., 2016). Cucumbers, cabbage and green olives account for the largest volume of vegetables and fruits commercially pickled in western countries while as raw mangoes, carrot, cauliflower and lemon is mostly used in the Asian countries including India. Pickling imparts unique characteristics to fruits and vegetables. Desirable changes in flavor, texture and color take place in fermented pickles (Barrett, 2003).

5.4 PRESERVATION BY SUGAR

Sugaring is a method of food preservation that requires the food to be dehydrated and then packed with either crystalline sugar or with liquids containing high amount of sugar such as syrup, honey or molasses. The basic concept of sugaring is that high concentration of sugar makes the food hypertonic in nature. Since hypertonic solution does not allow the growth of microbes that may otherwise diminish the nutritional value and quality of food (Kumar, 2019). Sugaring is commonly used to preserve fruits as well as vegetables such as ginger. Some of the most popular sugared food products include jams, jellies, concentrated soft drinks like orange squash etc. Sugaring mainly reduces the action of internal food enzymes by inhibiting the bacterial growth, resulting in the preservation of food from spoilage.

5.5 PRESERVATION BY LOW TEMPERATURE

Food spoilage either caused due to microorganism, improper packaging, lack of storage, malpractices followed during handling or due to lack of suitable preservation method possesses a great threat to the whole food supply chain (POPA *et al.*, 2019). So, preserving food using low temperatures has been considered efficient from the past. Chilling and freezing has been used for preserving commercial food products since 1875, with the invention of ammonia refrigeration system (Prokopov and Tanchev, 2007).

5.5.1 Refrigeration

Refrigeration is ubiquitary in our daily lives including homes and vehicles. Almost every food industry is dependent on refrigeration as an essential part of the food chain which is being used in all stages from processing to distribution, retail and final consumption at consumer's home (Tassou *et al.*, 2009). The excessive need for refrigeration at different levels has created a vast variety of refrigeration and cooling techniques (Zink *et al.*, 2010).

Refrigeration is artificial cooling which brings down the temperature below ambient (room) temperature where heat is transferred from low temperature reservoir to high temperature reservoir either mechanically or magnetically (Kumar, 2019). It is used to extend the shelf life of fresh (perishable) and processed foods by reducing the rate of biochemical and microbiological changes in it (Prokopov and Tanchev, 2007). Generally, temperature in the refrigerator is maintained between 4°C to 10°C but it could be used below 0 °C to -1 °C as well, for fresh meat and fish and their products (Morris, 1991). Application of low temperature retard growth of microbes or if grow, that to at a much slower rate. Enzymes catalytic rate in fresh food products such as raw meat, fish, fruits and vegetables also becomes much slower at low temperature further preventing its rotting and spoilage.

Microorganisms have different growth rate and generation time at different temperature ranges. The growth of microbes is maximum when optimum growing temperature conditions are available. Some microbes can even grow at very low temperatures (below 0°C) or at very high temperature, so necessity of recognising optimum temperature for microbial growth was there. Based on that, four broad categories of microorganisms depending on the temperature range for growth were thermophilic (minimum 30-40°C, optimum 55-65°C), mesophilic (minimum 5-10°C, optimum 30-40°C), psychrotrophic (minimum <0-5°C, optimum 20-30°C) and psychrophilic (minimum <0-5°C, optimum 12-18°C) (Mishra and Gamage, 2007).

Refrigeration could prevent the growth thermophilic and many mesophilic microorganisms but some pathogenic microbes have tendency to grow at prolonged storage below 5°C and may cause food poisoning if microbial population increases. For example *Listeria* spp., *Aeromonas hydrophilica* and *Yersinia enterocolitica*, etc. Enteropathogenic E. coli ($O_{157:H7}$) may cause haemorrhagic colitis after ingestion of as few as 10 cells therefore, some pathogenic species are sufficiently virulent to cause poisoning at fewer cell count (Mishra and Gamage, 2007).

5.5.2 Freezing

Freezing is another foremost method of food preservation which could inhibit the growth of microbes for longer time than refrigeration only under frozen condition. In freezing, the temperature of food product is reduced below its freezing point (-18 °C or less) and a proportion of the water present inside the food undergoes a change in its state to form ice crystals (Vaishali *et al.*, 2020).. The preservative effect is due to resulting concentration of dissolved solutes in unfrozen water after freezing, which leads to the reduction of pH values and water activity (Ray, 2000). However, freezing not only preserves the food product but original sensory and nutritional qualities are also maintained for longer period (Degner *et al.*, 2013). Ready-to-eat meals are frozen by quick freezing to a temperature near -15 to -20°C (Evan, 2008) using cryogenic technique.

Efficient freezing method is associated with the size of ice crystals formed in the food on freezing which has great influence on product quality and its texture. The smaller the size of ice crystals formed, the lesser damage it would cause to the cell structure of the food (Kennedy, 2000). For obtaining better and high quality product after thawing, quick freezing and storing at a constant sub-freezing temperature (Petzold *et al.*, 2009) is must. Apart from freezing technique, ice crystals formation, i.e. ice nucleation and its growth, is effected by characteristics of food being frozen (its mass and component) (Degner *et al.*, 2013)

Cooling can extend the storage life of fresh perishable fruits (1°C) and vegetables (4°C) only to several days but freezing (18 and -35°C, respectively) could increase there storage life to several weeks or months. There is need for proper selection of design and heat transfer mechanisms for storage of fruits and vegetables because they continue to respire and generate heat even after harvesting during storage. The quality of frozen products is heavily depends on the rate of freezing and velocity of refrigerated air which affects the rate of moisture loss from the products in addition to the rate of heat transfer (Fennema *et al.*, 1975).

With the advancement of technology, various new, energy saving, environment friendly refrigeration technologies have been invented replacing older ones have been invented. To reduce environmental impact, tested technologies based on air cycle refrigeration, ejector refrigeration system, sorption-adsorption refrigeration, stirling refrigeration cycle, thermoacoustic refrigeration, thermoelectric refrigeration or trigeneration, technologies that can reduce the quantity of used energy, that can use solar energy instead of conventional ones are developed (POPA *et al.*, 2019). The food industry still depends mainly on the vapor compression refrigeration cycle for food preservation and processing (Tassou *et al.*, 2009). Therefore, more research and developments are needed to replace the existing techniques at industrial level.

5.6 PRESERVATION OF FOOD BY HEAT TREATMENT

Heating of food to increase its shelf life has been long used to preserve food. Thermal treatment of food inactivates microorganism present in food commodity to accomplish either commercial sterilization or pasteurization. Pasteurization is thermal destruction of all viable forms of microorganisms present in the food and used to extend the limited shelf-life of food under refrigerated conditions. Sterilization of foods is normally done for canning, bottling or retorting to maintain the safety and wholesomeness of ready-to-eat foods for longer duration at room temperature and therefore, known as shelf-stable foods as it could be stored for a year. Blanching is a pretreatment or mild heat treatment done before processing or preserving food by inactivating enzymes and improving quality of food. All these thermal treatments differ widely with respect to the classification or type of microorganism targeted the range of temperatures that must be used, and the type of equipment systems capable of achieving such temperatures. However, these heat treatment processes must be carried out under a strict combination of temperature and time control to ensure the killing of pathogenic and non-pathogenic microorganisms (Teixeira, 2007).

5.6.1 Pasteurization

The name "pasteurization" was originally taken from the name of French scientist Louis Pasteur, who invented the process of heating liquids knowns as pasteurization (wine and beer) at a relatively mild temperature (about 55 C) for a short time to prevent spoilage which was later applied to milk to kill pathogenic microbes and reduce the number of spoilage organisms (Peng *et al.*, 2020). Thus, pasteurization refers to the process of heat treatment of food (usually below 100 C) to destroy micro-organisms of public health significance and kill relatively heat-sensitive microorganisms that cause spoilage, while maintaining with minimal changes in food (sensory and nutritional properties). Pasteurization do not destroy or kill all micro-organisms in foods but they target pathogens and lower levels of spoilage organisms that may grow during storage and distribution(Peng *et al.*, 2020). It is inefficient at killing the spore-bearing organisms, such as thermophilic *Bacillus subtilis*. Yeast is killed by heating at 60 C–65 C and the resistant mold spores, negative in most cases, are heated up to 80 C for 20 min (Prokopov and Tanchev, 2007).

There are several types of pasteurization techniques, one of which is batch pasteurization which is also called low temperature short time process. The fluid foods (milk or juice) are held in a tank where they are heated to 62.8 C for 30 min. A batch pasteurizer or vat pasteurizer consists of a steam jacket kettle or a tank equipped with steam coils in which juice or milk is heated to the desired temperature. There other type of pasteurizer is continuous pasteurization type this is also called high-temperature-short-time (HTST) process based pasteurization. Foods such as milk are subjected to 71.7 C for about 15 seconds or more by flowing through different heat exchangers. In continuous pasteurization, generally, tubular heat exchangers, plate heat exchangers, and scraped surface heat exchangers are used, depending on the viscosity of the fluid food material. The heating medium is usually steam or water (Fennema *et al.*, 1975).

Ultra-High-Temperature (UHT) equipment are much similar to HTST units. The controls are similar but the operating temperature of UHT are higher compared to HTST. Generally, in the UHT, the holding time is 3 second. A UHT treatment is desired because of its greater bacterial destruction capability or its beneficial effects on treated food product (Ramesh, 2007).

5.6.2 Canning

Thermal processing or canning consists of heating food in containers or retorts (autoclaves or 'pressure cookers') at specified temperatures for predefined length of time. These processing time are calculated on the basis of heat resistance and microbial lethality causing bacterial inactivation in canned food for assuring food safety to the consumer. Canning employs high heat treatment temperature which could affect the nutritional and overall quality of food. Some heat-sensitive vitamins get degraded and other quality factors are also reduced which is undesirable. So, highly precise calculations are done before canning of food.

Canning can be done both in batch and continuous systems. As the name suggest, batch systems are made up of individual batch retorts that operate intermittently. Scheduling is done skilfully so that workers could move from retort to retort, manually unloading and reloading each retort as its scheduled process cycle comes to an end. In continuous systems, conveyor belts automatically move the cans and the food is fed into and out of retort systems that operate automatically and continuously by machine over one or more working shifts. The continuous systems are monitored carefully to prevent any lagging behind or misfunctioning of machine controls (Ramesh, 2007).

5.6.2.1 Batch systems

Batch systems are huge pressure vessels in which containers, baskets or crates of food are sterilized with saturated steam or water under pressure. These retorts can be either vertical or horizontal in configuration. The vertical type batch retort system is marked above all batch retorts as every food science pilot plant or laboratory have this system. A typical production vertical unit will measure nearly 4ft in diameter by 8 or 9ft in height. Cans are loaded in baskets that are connected by chain hoist for lifting and lowering into the vertical retort. In the case of horizontal retorts, cans are loaded into wheeled carts that roll on rails into and out of the retorts. Most retorts systems are designed to hold either three or four crates or carts, with a total capacity of more than 2000 pint-size cans or 400 gallon-size cans (Teixeira, 2007).

The basic design of these retorts has not changed since twentieth century because of the simplicity of their design and operation and their versatility to accommodate virtually all can sizes and shapes. The filled and sealed unsterilized cans enter the cook room continuously from the filling line operations and fully processed sterilized cans leave the cook room continuously. These operations are carefully monitored by the operators to ensure that the scheduled process is delivered for each batch. These batch retorts are designed for flexible or semi-rigid retortable packaging systems which require overriding air pressure to protect packages from bursting during processing. These operate with water spray, water cascade, or steam–air mixtures with overriding air pressure, and are capable of delivering end-over-end agitation when desired (Ramesh, 2007).

5.6.2.2. Continuous retort systems

Continuous retort systems fill and seal containers automatically and continuously which move from atmospheric conditions into a pressurized steam environment, held or conveyed through that environment for the specified process time, and then returned to atmospheric conditions for further down-stream handling operations. The best-known commercially available systems that accomplish these requirements are the crateless retort, the continuous rotary cooker, and the hydrostatic sterilizer (Teixeira, 2007).

5.6.3. Blanching

Blanching is a pretreatment process or relatively mild heat treatment given to raw foods before processing or preservation processes. It is the step involved prior to preservation techniques such as freezing, frying, drying and canning accomplished after washing, sorting, cutting, chopping, dicing or slicing of the raw food (Arroqui, 2003). It is often required for releasing entrapped air in the tissue of raw materials, inactivate enzymes, modifying texture, preserving colour, flavor and nutritional value which influence food quality. Vegetables and fruits are rapidly heated to a predetermined temperature and maintained at a specified amount of time, mainly 1 to less than 10 min. Then the blanched product is rapidly cooled or passed immediately to a next process. The blanching time for any food depends on the time required for inactivation of peroxidase and polyphenoloxidase enzymes present in fresh food products (Xiao *et al.*, 2017).

Parboiling in case of raw paddy is also a form of blanching in hot water and steam (Miah, 2002). In the context of food preparation in cooking, blanching is an alternative term for par boiling. In the food processing industry, blanching is done by drawing food material to travel on a conveyor through an atmospheric steam tunnel or hot water bath then the food is plunged into boiling water for few minutes and immersed quickly in cold water (shocked) to immediately halt further cooking due to heat. Food is often blanched to soften it, or to partly or fully cook it, or to remove a strong undesirable taste (for example, of bacon, cabbage, or onions) (Xiao *et al.*, 2017).

Blanching is done in various form depending upon the medium used for heating. Water blanching is performed in hot water at temperatures ranging typically from 70°C to 100°C. There are water blanchers with a screw or chain conveyor to transport the product inside the tank and others use a rotary drum to immerse and convey the product where hot water is present (De Corcuera, 2004). Steam blanching performed by conveying through a chamber where food-grade steam at approximately 100°C is directly injected. Steam blanching is usually used for cut and small sized products and which requires lesser time than water blanching because the heat transfer coefficient of condensing steam is greater than that of hot water whereas smaller size of food product enable larger surface area for contact with steam (Teixeira, 2007).

5.7 PRESERVATION BY CONTROLLED ATMOSPHERE AND MODIFIED ATMOSPHERE

Controlled atmosphere and modified atmosphere are conditions in which addition or removal of gases from storage rooms, transportation containers or packages is done to change the levels of gases present around the food such as oxygen, carbon dioxide, nitrogen and ethylene etc., (Floros and Matsos. 2005). The maintenance of a constant gas phase is difficult to achieve but modification of the atmosphere is used mainly for larger storage of fresh and partly processed food (meat, fish, fruits and vegetables) though individual packs are often flushed with gas composition around the product (usually in storage rooms or transportation containers) which is continuously monitored and controlled (Majidi *et al.*, 2014).

5.7.1. Preservation by Controlled atmosphere

In controlled atmosphere packaging, the gas composition is altered according to the product requirements inside the packaging which is monitored and maintained at a recommended level by means of scrubbers and the inlet of gases provided. Active packaging is a type which could provide a solution by providing materials that would absorb or release a specific compound in the gas phase. Compounds that can be absorbed are carbon dioxide, oxygen, water vapor, ethylene, or volatiles that influence taste and aroma. (Gorris and Peppelenbos, 2007). In controlled atmosphere storage, addition to the advantage of gas alteration facility for any food product, refrigeration facility is also available which helps in increase of shelf life for longer duration in comparison to modified gas composition only.

5.7.1.1 Alteration of gas composition

Management of gas composition in controlled atmosphere storage is very important but controlling depletion and evolution of gases in the system to withstand product requirement is marked essential. Earlier the level of CO_2 released in the system was roughly equal to the extent of O_2 depletion in accordance with the respiratory equation. The removal of the respired carbon dioxide was important to prevent injury to pear and apple cultivars which have sensitivity towards CO_2. Therefore, scrubbing of respired CO_2 was earlier accomplished by using sodium hydroxide (caustic soda), but currently activated carbon and hydrated lime are well known to serve this purpose. Water scrubbing and oxygen purging technologies are also being included in these systems (Ben-Yehoshua *et al.*, 2005).

Replacing oxygen with liquid nitrogen was common means to establish low O_2 atmospheres. Two technological advances made had greater impact on controlling atmosphere. The air separator technologies of pressure swing adsorption (PSA) and hollow fiber membrane air separation which performed the same function but these operated on entirely different principles (Floros and Matsos, 2005). The pressure swing absorption system used a molecular sieve for adsorbing oxygen due to alteration in pressures, it could direct the nitrogen enriched gas stream containing depleted oxygen level into the room and later exhaust the oxygen to the atmosphere (Majidi *et al.*, 2014).

The hollow fiber membrane system works on the permeability of the fiber walls which are selectively permeable to nitrogen and oxygen gas. These gases are forced through the length of fibers by moving in and out to the chamber or waste gas stream under high pressure and temperature. Nitrogen gas passes slowly through the wall of the fiber compared to oxygen, so that O_2 has passed before reaching the farther end of the fiber through the wall, leaving nitrogen in the output stream. Similarly CO_2 and ethylene gases could also be separated from nitrogen with this system, raising the chance of using it as scrubber for these gases. Both the systems are compatible with computer based sensing and control systems (Floros and Matsos, 2005).

But there are more improved systems on which current research is being carried out for its proper implementation. Dynamic control systems (DCSs) continually adjust the storage atmosphere, rather than maintaining a set point as in the conventional systems. it could use the product as sensor to judge the stressful conditions in the atmosphere. In the Netherlands, a strategy has been developed for the detection of the fermentation point of fruit crops in controlled atmosphere storage systems by using ethanol sensors which detected the increase in ethanol production as a result of change in the oxygen and carbon dioxide concentrations in the atmosphere and readjusted to a less stressful level (Gorris and Peppelenbos, 2007).

5.7.2. Preservation by Modified atmosphere

Preservation by modifying product atmosphere has been long used. The respiratory activity of the various plant products generates a low oxygen and high carbon dioxide concentration atmosphere, which itself retards the ripening of fruit. Modified atmosphere packaging is a preservation technique that minimize the physiological and microbial decay of food (perishable produce) by surrounding them with an atmosphere that is different from the normal composition of air. The gas composition and the technique depends on the types of produce and its requirement. In modified atmosphere packaging (termed "passive atmosphere"), the gas composition within the package is not monitored or adjusted with time as it passively occurring with no control (Gorris and Peppelenbos, 2007). However, under such conditions, there are chances of reduction and changes in some nutrients such as vitamins and ascorbic acid. Crop grains are preserved using carbon dioxide gas. Carbon dioxide prevents insects, molds and oxidation from damaging the grains. Nitrogen gas is also used grains for preservation and it also kills insect. During storage, gas content of CO_2, O_2 and N_2 may subsequently change as a consequence of packacage permeability, biological activities of packed product, chemical reaction, for example, of oxygen with some components of the foods like vitamin C (Prokopov and Tanchev, 2007).

The main distinguishing feature between modified atmosphere and controlled atmosphere is that, active involvement of humans stops at the moment of sealing of the food product. Modified atmosphere packaging is basically applied to smaller quantities of product but controlled atmosphere ranges from a few grams for herbs or spices to several kilograms for blueberries (Ben-Yehoshua *et al.*, 2005).

5.8 PRESERVATION OF FOOD BY FERMENTATION

Fermentation is the process of conversion or breakdown of carbohydrates (complex) into simpler substrates either under anaerobic or aerobic conditions by microorganisms which cause changes in the sensory (flavour, odour, etc.) and functional properties of food. Fermentation has wide applications in producing diversified and highly palatable food products from same raw ingredient (Melini *et al*, 2019). Microorganisms involved in the fermentation are namely bacteria, yeasts, and mycelial fungi and their enzymes (Potter and Hotchkiss, 2012).

Microorganisms are omnipresent so, microflora may be already present on the food substrate, or they may be present in or on the ingredients and utensils, or in the environment. But intentional fermentation is done by adding starter culture in the food like curd is prepared from its previous curd starter culture added into the milk. Basically, a suitable substrate, appropriate microorganism(s) and proper environmental conditions, such as temperature, pH, and moisture content must be available in order to initiate food fermentation (Joshi, 2016).

Fermentation improves the nutritional quality of raw ingredient used as substrate by increasing health benefits from the food product. Indigenous fermented foods such as dahi bhalle, papad, idli, and dosa, traditionally made at homes from locally available knowledge and raw materials, is now taken a step forward by entering the commercial lines. These local indigenous foods are now globally, using a wide range of different raw materials, microorganisms, and manufacturing techniques available due to industrial production of these food items and there longer self life (Joshi, 2016).

Fermented foods in table 1 are among the oldest processed foods and have formed a traditional part of the diet in almost all the countries. Fermentation done in combination with salting, cooking, smoking and sun drying is the techniques developed by ancient cultures to extend storage life of food. They form the major sector of food processing industry, including baked products, alcoholic beverages, dairy fermented products and soy sauce, etc (Rezac *et al.*, 2018). During the fermentations, the controlled action of selected microorganism is used to alter the texture of food, preserve food, production of acids and alcohols, or to improve flavours and aroma which increases the quality and value of raw material. Today, the preservative effect is supplemented by other unit operations in the industry (Melini *et al.*, 2019).

Table 1. Common fermented food

Fermented food	Raw ingredient	Fermenting microbe(s)
Beer	Barley	Yeasts
Wine	Grapes	Yeasts
Cider	Apples	Yeasts
Bread and baked products	Wheat	Yeasts
Yoghurt	Milk	Lactic Acid Bacteria
Cheese	Milk	Lactic Acid Bacteria
Tempeh	Soyabeans	Molds
Pickles	Cucumber, olives, etc	LAB and yeasts
Sauerkraut	Cabbages	LAB and yeasts

Source: Guizani, N. and Mothershaw, A. 2007. Fermentation as a Method for Food Preservation. Rahman, M. S. (ed). *Handbook of Food Preservation (Second Edition), Vol-1*. **CRC press, London, New York.**

5.8.1 Health benefits from fermentation

Various food material are fermented to get a new product, improved flavor, taste, aroma and other benefits but fermentation also impart health benefits (Hasan *et al.*, 2014). Fermentation of milks is done from past to obtain various products like curd, cheese and yoghurt by choosing specific microbial culture to prepare these different products from same ingredient (milk). Fermented milk contains Lactic Acid Bacteria (LAB), Bifidobacteria, and yeasts which grow in a protocooperative relationship. LAB includes several members of Lactobacillus, Leuconostoc, Lactococcus, Streptococcus, and Pediococcus genera which provide fermented milk its peculiar favour, texture, and nutritional value. Fermentation have a crucial role in the formation of bioactive components, which especially impart antioxidant, anti-hypertensive, anti-diabetic, and anti-allergic potential to the raw material (Marco *et al.*, 2017).

5.8.2 Preservative effect of Microbial Fermentation:

Raw vegetables under anaerobic condition undergo pickling because of the presence of naturally occurring bacteria which produce fermentation and increase its shelflife. Mainly for the pickling of vegetables they are submerged in brine (2-6%), spices or oil. Pickling increases the shelf life of the perishable products. Mainy fruits and vegetables are seasonal in nature and mainly lost in market glut or surplus. Thereby, fermenting these excessive surplus perishable products, can increase its value as well as help in availing the delight of the fruit or vegetable in new form of product. In tropical countries, cereals and root crops are fermented to a range of beverages and staple food products. Here fermentation impart flavor and aroma to the product and help in the formation of a whole new product. Fermented alcoholic beverages from cereals are whiskey, beer and ales are very much and has long history back then etc.

5.9 PRESERVATION OF FOOD BY LOWERING WATER ACTIVITY (a_w)

Water activity is one of the important factor in limiting the growth of microorganisms in food media by providing a preservative action. Water activity could be reduced by partial removal of the water (drying, reverse osmosis, concentration) or by adding substances which increase the osmotic pressure of the food or media such as sugars, ethanol, glycerol, salts, etc. (Prokopov and Tanchev, 2007).

Water activity is the ratio of vapor pressure of water and the vapor pressure of pure water in a system at the same temperature, or it could be defined as equilibrium relative humidity of the air surrounding the system at the same temperature. Water activity in the food products could be determined or estimated by various methods which have been reported in literature which includes equilibrium sorption rate method (isopiestic method), vapor pressure measurement method and hygrometric instrument method. In addition, freezing point (thermodynamic property) of food components could be used to measure water activity (Rizvi, 1995).

Of all the factors (oxygen, temperature, acidity, nutrient availability and pH, presence of natural, or added inhibitors, etc.) affecting microbial growth, death and survival in food products, the influence of water activity on vegetative growth in microorganisms and spores is one of the most complex and extensively studied by food microbiologists (Sereno *et al.*, 2001). Osmotic stress possessing adverse effect on microorganisms can hinder the sporulation process in spore-forming microorganisms, but bacterial endospores and some fungal spores have special requirements, such as optimum water activity values for initiating germination (minimum aw for germination is usually higher than the minimum aw for sporulation). In addition, secondary metabolites (toxins) are also affected by optimum water activity requirements. Therefore, germination, sporulation and toxin production are affected by water activity along with other environmental factors (Beuchat, 2002).

Lowering of water activity is done by following methods:

- Removing or separating out water: Drying, concentration and dewatering by centrifuge. Other unit operations such as baking, extrusion and frying also lower the water activity of food product,
- Addition of solutes: Addition of sugars to fruits and addition of salt to meat and fish could lower the water activity by creating osmotic differences.

Majority of microorganisms could survive in the available water present in their immediate environment and there metabolic activity could remain active in a narrow range of high water activities (Table 2). The most water activity (aw) tolerant microbial species are able to grow when osmotic pressure is as high as about 800 MPa and they could grow slowly below 0.62 aw units (Sereno *et al.*, 2001).

Table 2. Minimum aw of food borne pathogens at optimum laboratory conditions

Food Borne Pathogen	Minimum a_w
Staphylococcus aureus (aerobic)	0.86
Staphylococcus aureus (anaerobic)	0.91
Listeria momocytogenes	0.92
Bacillus cereus	0.93
Escherichia coli	0.93
Clostridium boyulinum A and B	0.94
Shigella app.	0.96
Aeromonas hydrophilia	0.97
Campylobacter jejuni	0.99

Source : Gorris, L. G. M. and Peppelenbos, H. W. Modified-Atmosphere Packaging of Produce. Rahman, M. S. (ed). *Handbook of Food Preservation (Second Edition), Vol-1.* **CRC press, London, New York.**

The presence of solute exerts additional affect on potential growth of microorganisms. Ionic solutes (NaCl and KCl) have more potential to inhibit microbial growth than non-ionic solutes such as sugars. Polar solutes such as glycerol, unlike salts and sugars, rapidly permeate most bacteria but not yeast, e.g. *Saccharomyces ronxii* and *Debaryomyces hanseni*. However, this could not always the same for all microbial species as some species could survive in extreme conditions as well, for example, *Staphylococcus aureus* is extremely salt-tolerant and more sensitive at a higher aw in glycerol than in sodium chloride. Lowering the water activity (aw) by different methods influence the rate of enzymatic and chemical changes in foods. Whereas, most of the microbiological growth is completely stopped below aw = 0.6, but some deteriorative enzymatic reactions could still continue that causes food spoilage, such as lipid oxidation (Prokopov and Tanchev, 2007).

Conclusion

One of the major revolutionary inventions of human civilization was acquiring the knowledge to preserve foods. Foods being organic perishable substances deteriorate in quality due to a wide range of reactions including physical, chemical, enzymatic and microbiological. These types of spoilage can be prevented to a greater extent by numerous preservation techniques, mostly by slowing or inhibiting the microbial growth and enzymatic activity. With time different preservation techniques had been evolved to preserve foods and to maintain their nutrition value and texture. However, the preservation method used should be suitable for the food condition because not all these preservation techniques are able to maintain the freshness and organoleptic properties of the resulting food products.

REFERENCES

- Agrahar-Murugkar, D. and Jha, K. 2010. Effect of drying on nutritional and functional quality and electrophoretic pattern of soyflour from sprouted soybean (Glycine max). J Food Sci Technol. 47(5): 482–7.
- Ahmed, J. and Ramaswamy, H. S. 2004. Microwave Pasteurization and Sterilization of Foods. In: *Food Science and Technology*. pp 691-709, New York, Dekker.
- Arroqui, C., Lopez, A., Esnoz, A. and Virseda, P. 2003. Mathematical model of heat transfer and enzyme inactivation in an integrated blancher cooler. *Journal of Food Engineering*, **58**:215–25.
- Barnwal P, et al. 2010. Effect of moisture content and residence time on dehulling of flaxseed. J Food Sci Technol., 47(6): 662–7.
- Baron, F., and Gautier, M., 2016. Microbial spoilage. Handbook of Food Science and Technology 1: Food Alteration and Food Quality. Wiley, New York, NY, pp. 55_97.
- Ben-Yehoshua, S., Beaudry, R. M., Fishman, S., Jayanty, S. and Mir, N., 2005. Modified atmosphere packaging and controlled atmosphere storage. Environmentally friendly technologies for agriculture produce quality. Ben-Yehoshua, S. (ed), Taylor and Francis Group LLC, Boca Raton, USA.
- Beuchat, L.R. 2002. *Water activity and microbial stability*. Fundamentals of Water Activity, pp 14-15. IFT Continuing Education Committee, Anaheim, California.
- Bonazzi, C. and Dumoulin, E. 2011. Quality Changes in Food Materials as Influenced by Drying Processes. Modern Drying Technology Volume 3: Product Quality and Formulation, First Edition, Published 2011 by Wiley-VCH Verlag GmbH & Co. KGaA.
- Coronel, P., Simunovic, J. and Sandeep, K. P., Temperature profiles within milk after heating in a continuous-flow tubular microwave system operating at 915 MHz. *Journal of Food Science*, **68**(6):1976-1981.
- Davidson, P. M. and Critzer, F.M., 2012. Interventions to inhibit or inactivate bacterial pathogens in foods. In: Oyarzabal, O.A., Backert, S. (Eds.), Microbial Food Safety: An Introduction. Springer, New York, NY, pp. 189_202.

- De Corcuera, J. I. R., Cavalieri, R. P. and Powers, J. R. 2004. Blanching of Foods *Encyclopedia of Agricultural, Food, and Biological Engineering*. pp 1-5. Marcel, Dekker, New York.
- Degner, B. M., Olson, K. M., Rose, D., Schlegel, V., Hutkins, R. and McClements, D. J. 2013. Influence of freezing rate variation on the microstructure and physicochemical properties of food emulsions. *Journal of Food Engineering*, **119**:244–253.
- Desrosier, N. W. and Singh, R. P. Food preservation. Encyclopaedia Britannica Inc.; 2014. https://www.britannica.com/topic/food-preservation
- Doyle, M. P. 2009. Compendium of the microbiological spoilage of foods and beverages. Food microbiology and food safety. New York: Springer.
- Enfors S-O. 2008. Food microbiology. Stockholm: KTH-Biotechnology; 2008.
- Erkmen, O. and Bozoglu, T. F. 2016. Enzymatic and Non-enzymatic Food Spoilage *Food Microbiology: Principles into Practice,* First Edition. Published 2016 by John Wiley & Sons, Ltd.
- Evan, J. A. 2008. *Frozen Food Science and Technology*. Wiley Online Library.
- Fabunmi, O. A., Osunde, Z. D., Alabadan, B. A. and Jigam, A. A. 2015. Influence of moisture content and temperature interaction on mechanical properties of DESMA (Novella pentadesma) SEED. J Adv Food Sci Technol., 2(2): 81–5.
- Fellows, P. J. 2009. Food processing technology: principles and practice. 3rd ed. Cambridge: Woodhead Publishing.
- Fennema, O. R, Karel, M., and Lund, D. B. 1975. Freezing preservation. In Fennema, O. R. (ed.), *Principles of food sciences part II. Physical principles of food preservation*. pp 173-215. Marcel, Dekker, UK.
- Floros, J. D. and Matsos, K. I. 2005. Introduction to modified atmosphere packaging. In: Han, J. H. *Innovations in Food Packaging*. pp 159-171. Elsevier Academic Press. California, UK.
- Gorris, L. G. M. and Peppelenbos, H. W. 2007. Modified-Atmosphere Packaging of Produce. In: McElhatton, A. and Marshall, R. J. (eds). *Food Safety A Practical and Case Study Approach*. pp 316-329. Springer Science + Business Media, LLC, London, UK.
- Gould, G. W. 2000. Preservation: past, present and future. *Brrtish Medical Bulletin* 2000, 56:1, 84-96
- Guine, R. P. F. 2018. The drying of foods and its effect on the physical-chemical, sensorial and nutritional properties. International journal of food engineering, 4(2): 93-100.
- Hasan, M. N., Sultan, M. Z. and Mar-E-Um, M. 2014. Significance of Fermented Food in Nutrition and Food Science. *Journal* OF *Scientific Research*, 6(2): 373-386.

- Jangam, S. V., Law, C. L., Mjumder, A. S. 2010. Drying of foods, vegetables and fruits, vol. 1, 1st ed. Singapore; 2010.
- Jay, J. M. 2000. Modern food microbiology. 6th ed. Gaithersburg: Aspen Publishers; 2000
- Joshi, V.K. 2016. *Indigenous Fermented Foods of South Asia*, 1st edition. pp 886. CRC Press, Boca Raton, USA.
- Kennedy, C. J. 2000. Future trends in frozen foods. In: Kennedy, C. J. (ed.), *Managing Frozen Foods*. pp 263-278. Woodhead Publishing Limited, Cambridge.
- Kreyenschmidt, J. and Ibald, R. 2012. Modeling shelf life using microbial indicators. In: Nicoli, M.C. (Ed.), Shelf Life Assessment of Food. CRC Press, Boca Raton, FL, pp. 127_168.
- Kumar, A. 2019. Food Preservation: Traditional and Modern Techniques. *Acta Scientific Nutritional Health*, **3**(12): 45-49.
- Kumar, A. 2019. Food Preservation: Traditional and Modern Techniques. *Acta Scientific Nutritional Health* 3.12 (2019): 45-49.
- Levine, H. and Slade, L. 1988. Principles of "cryostabilization" technology from structure/property relationships of carbohydrate/water systems—a review. Cryo Lett., 9(21): 21–63.
- Majidi, H., Minaei, S., Almassi, M. and Mostofi, Y. 2014. Tomato quality in controlled atmosphere storage, modified atmosphere packaging and cold storage. *Journal of Food Science and Technology*, **51**(9): 2155-2161.
- Marco, M. L., Heeney, D.; Binda, S.; Cifelli, C.J.; Cotter, P.D.; Foligné, B.; Gänzle, M.; Kort, R.; Pasin, G.; Pihlanto,A.;etal. HealthBene tsofFermentedFoods: MicrobiotaandBeyond. Curr. Opin. Biotechnol. 2017, 44, 94–102.
- Melini, F., Melini, V., Luziatelli, F., Ficca, A. G. and Ruzzi, M. 2019. Health-Promoting Components in Fermented Foods: An up-to-date Systematic Review. *Nutrients*, **11**: 118
- Miah, M. A. K., Haque, A., Douglass, M. P. and Clarke, B. 2002. Parboiling of rice. Part II: Effect of hot soaking time on the degree of starch gelatinization *International Journal of Food Science and Technology*, **37:** 539–545.
- Mishra, V. K. and Gamage, T. V 2007. Postharvest handling and treatments of fruits and vegetables. Rahman, M. S. (eds). *Handbook of Food Preservation* (2nd edition*), Vol-1*. pp 49-72. CRC press, London New York.
- Modi, H. A. 2009. An introduction to microbial spoilage of foods. In: Modi, H.A. (Ed.), Microbial Spoilage of Foods. Aavishkar Publishers, Jaipur, India, pp. 43_66.
- Morris, C.E. 1991. Forging new links in the cold chain. *Food Engineering*, **7**: 61.

- Nummer, B. A. 2002. Historical origins of food preservation. http://nchfp.uga.edu/publications/nchfp/factsheets/food_pres_hist.html
- Pehrsson, P. R. and Haytowitz, D. B. 2016. Food composition databases. Encyclopedia of food and health, 16-21.
- Petruzzi, L., Corbo, M. R., Sinigaglia, M. and Bevilacqua, A. 2017. Microbial Spoilage of Foods: Fundamentals. The Microbiological Quality of Food. DOI: http://dx.doi.org/10.1016/B978-0-08-100502-6.00002-9
- Petzold, G. and Aguilera, J. M. 2009. Ice morphology: fundamentals and technological applications in foods. *Food Biophys*, **4**:378–396.
- POPA, E. E., Mitelu , A. C. and POPA, M. E. 2019. *Trends in Refrigeration Technologies used for Food Preservation – A review*. Scientific Bulletin 25, Series F. Biotechnologies. pp 205-210.
- Potter, N. N. and Hotchkiss, J. H. 2012. *Food Science*. Springer Science and Business Media, LLC, London, UK.
- Prokopov, T. and Tanchev, S. 2007. Methods of Food Preservation. In: McElhatton, A. and Marshall, R. J. (eds). *Food Safety A Practical and Case Study Approach*. pp 3-25. Springer Science and Business Media, LLC, London, UK.
- Prokopov, T. and Tanchev, S. 2007. Methods of Food Preservation. Food Safety A Practical and Case Study Approach. Springer, ISBN-10: 0-387-33509-9.
- Rahman, M. S. 2007. Food Preservation: Overview Handbook of Food Preservation. Rahman, M. S. (eds). *Handbook of Food Preservation (2nd edition), Vol-1*. pp 3-18. CRC press, London, New York.
- Rahman, M. S. 2007. Food preservation: Overview. Handbook of food preservation. CRC Press, ISBN: 10: 1-57444-606-1
- Ramesh, M. N. 2007. Pasteurization and Food Preservation. In: McElhatton, A. and Marshall, R. J. (eds). *Food Safety A Practical and Case Study Approach*. pp 3-25. Springer Science + Business Media, LLC, London, UK.
- Ray, B. 2000. *Fundamental Food Microbiology*. 2nd Edition. CRC Press, London.
- Rezac, S., Kok, C.R., Heermann, M. and Hutkins, R. 2018. Fermented Foods as a Dietary Source of Live Organisms. *Frontiers in Microbiology*, **9**: 1785.
- Rizvi, S. S. H. 1995. Thermodynamic properties of foods in dehydration. In: Rao, M. A. and Rizvi, S. S. H. (eds) *Engineering Properties of Foods*. 2nd edition. Marcel Dekker, New York.
- Roos, Y. and Karel, M. 1991. Plasticizing effect of water on thermal behavior and crystallization of amorphous food models. J Food Sci., 56(1): 38–43.
- Sadat Kamal Amit†, Md. Mezbah Uddin†, Rizwanur Rahman, S. M. Rezwanul Islam and Mohidus Samad Kha. 2017. A review on mechanisms and commercial aspects of food preservation and processing. *Agric & Food Secur, 6:51, 1-33*

- Sereno, A. M. Hubinger, M. D. Comesana, J. F. and Correa, A. 2001. Prediction of water activity of osmotic solutions, Journal of Food Engineering, **49(**2): 103-114.
- Srivastava, R. P. and Kumar, S. 2003. Fruit and vegetable preservation: principles and practices, International book distributing company, Lucknow, U.P. India.
- Steele R. Understanding and measuring the shelf-life of food, 1st ed. Woodhead Publishing Limited; 2004.
- Tassou, S. A., Lewis, J. S., Ge, Y. T., Hadawey, A. and Chaer, I. 2009. A review of emerging technologies for food refrigeration applications. *Applied Thermal Engineering*, **30**(4): 263.
- Teixeira, A. A. 2007. Thermal processing of canned foods. In: Heldman, D. R. and Lund, D. B. (eds), *Food Engineering Handbook*, 2nd edn. pp. 592–659. CRC Press, Taylor & Francis Group, Boca Raton, Florida, USA.
- Vaishali, Sharma, H. P., Dholu, U., Sharma, S. and Patel, A. 2020. Effect of freezing systems and storage temperatures on overall quality of perishable food commodities. *The Pharma Innovation Journal*, **9**(9): 114-122.
- White, G. W. and Cakebread, S. H. 1966. The glassy state in certain sugar-containing food products. Int J Food Sci Technol., 1:73–82.
- Xiao, H. W., Pan, Z., Deng,L. Z., Hamed, M., Mashad, E., Yang, X. H., Mujumdar, A. S., Gao, Z. J. and Zhang, Q. 2017. Recent developments and trends in thermal blanching – A comprehensive review. *Information Processing in Agriculture*, **4**: 101–127.
- Zink, F., Vipperman, J. S. and Schaefer, L. A. 2010. Environmental motivation to switch to thermoacoustic refrigeration. *Applied Thermal Engineering*, **30:** 119–126.

3

CHAPTER

INNOVATIVE TECHNOLOGIES FOR FOOD PRESERVATION

Rukhsana Rahman*, Neeraj Gupta, Meenakshi Trilokia, Fozia Hameed, Nadira Anjum and Anjum Ayoub

Division of Food Science and Technology, Sher-e-Kashmir University of Agricultural Science and Technology-Jammu, Chatha-180009, India.

*Corresponding author:*rukhsananrahman786 gmail.com*

*Corresponding Author: *[1]s.fozia011@gmail.com*

ABSTRACT

Several techniques have been developed during the 20th century in order to preserve foods. These innovative technologies vary considerably and embrace physical technologies (e.g., high hydrostatic pressure and high-pressure homogenization), electromagnetic technologies (e.g., pulsed electric fields, Ohmic heating, microwaves, radio-frequency, infrared heating and UV-light), acoustic technologies (e.g., ultrasound) and cold plasma treatment. In this chapter, the theoretical background and definition of the technologies are explained together with a description of the equipment, main technological/processing parameters, and some applications from a technological point of view.

Key words: preservation, food processing, quality, innovative techniques

1. INTRODUCTION

Since ancient times, the approach of the food industry was to provide safe food product with long shelf-life; however, presently it is not enough to simply produce safer food as consumers demand products with a significant nutritional contribution, bioactive compounds, and good sensory properties (Hameed et al., 2018). The important food quality attributes, such as taste, texture, appearance, and nutritional value, are strongly dependent on the way food is processed (Knoerzer *et al.*, 2016). Microorganisms are the main target organisms for food spoilage and poisoning so are targeted by different food preservation procedures. Food processing methods used by industry rely either on microbial inactivation or inhibition of microbial growth. Conventional heat-dependent pathogen-reduction methods such as thermization, pasteurization and in-container sterilization can adversely affect taste, nutritional value and appearance (Hameed et al., 2018). In recent years, the food industry has shown increased interest in innovative approaches which can provide products of proven quality and with a reduced environmental cost, increasing the products added value (Pereira and Vicente, 2010). Innovative technologies that take advantage of physical forces and phenomena such as high hydrostatic pressure, electric and electromagnetic fields, and pressure waves, for example, high pressure processing (also in combination with heat), microwave processing, ohmic heating, pulsed electric field (PEF) processing, ultrasound processing (liquid and airborne), and ultraviolet light (UV) processing (Canovas *et al.*, 2011). Among them, (1) physical technologies (e.g., high hydrostatic pressure (HHP) and high-pressure homogenization (HPH)); (2) electromagnetic technologies (e.g., pulsed electric fields (PEFs), high-voltage electrical discharges, ohmic heating (OH), chemical electrolysis, microwaves (MWs), radio-frequency (RF), cold plasma, UV-light (continuous and pulsed)), and oscillating magnetic fields; (3) acoustic technologies (e.g., ultrasound (US) and shockwaves (SWs) and (4) others such as membrane filtration (MF) and dense phase CO2 (DPC) have been extensively studied.

2. INNOVATIVE FOOD PRESERVATION TECHNOLOGIES

The purpose of using these technologies is to inactivate the activity of the microorganisms present in the food and also certain enzymes of interest without destroying the nutritional and sensory components that are normally affected during heat treatment. Innovative processes are therefore being developed as an alternative to traditional thermal methods (Barba *et al.*, 2015). Some of the innovative technologies used for food preservation are described below.

2.1 Pulsed Electric Field (PEF)

Over the last few years, researchers have been paying more attention on PEF. It is one of the innovative technology for food preservation that uses short bursts of electricity for microbial inactivation and causes minimal or no detrimental effect on food quality attributes. Thus, PEF can be extensively used for the processing of liquid and semi-liquid food products along with selected solid foods. (Barba *et al.*, 2015).

Principle: The basic principle of the PEF technology is the application of high-voltage pulses to foods placed in between two electrodes. The duration of pulse is in the range of micro to milliseconds and intensity in the order of 10-80 kV/cm. (Puertolas and Barba, 2016). In general, pulsed electrical currents delivered to a food product placed in between a set of electrodes. The applied high voltage produces electric field may lead to microbial inactivation. It is important to keep the treatment temperature as low as possible so, as to avoid heat damage to the treated food product and deter off-flavours. (Mathavi *et al.*, 2013). It has been suggested that for orange juice, PEF treatment at 35 kV/cm for 590 ms was good in comparison to heat pasteurization at 94.6 C for 30 seconds. (Yeom *et al.*, 2000).

To ensure the safety of a food product, certain factors such as intensity of the electric field, the pulse shape, width and frequency, the total treatment time, the electrode configuration, and the temperature needs to be considered for an effective process. Other factors distressed the performance of the PEF process includes type, concentration, and growth stage of microorganism to be inactivated and pH, antimicrobials, and ionic compounds, electric conductivity, and medium ionic strength of the treatment media.

Pulse Electric Field System: It consist of different components: Power supply (a high voltage power source), an energy storage capacitor bank, a charging current limiting resistor, A switch (to discharge energy from the capacitor across the food and a treatment chamber), A pulse shaping and triggering circuit in some cases, a treatment chamber, A pump: for the supply of feed of product to the chamber and a cooling system (for control of the feed and output material temperature).

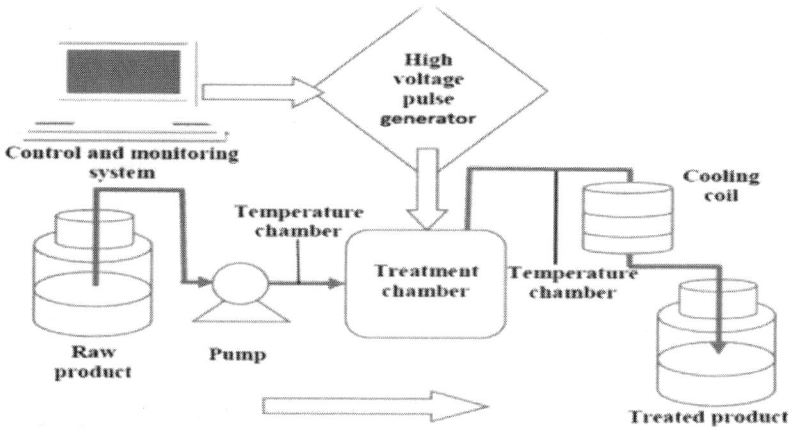

Fig. 1: Pulse Electric Field Set –up

Mechanism:- There are two main mechanisms that explain the microbial inactivation by PEF. a) Electroporation: in this phenomenon, cell exposed to high voltage electric field pulses temporarily weakens the lipid bilayer and proteins of cell membranes. The impact of an electric field on a microorganism cell is to hike permeability of the membrane owing to membrane compression and poration. B) Electrical breakdown: Destruction of cell due to electrical breakdown of the cell membrane instigated by the application of high-voltage pulses. Zimmermann (1986).

Applications

PEF is applied in processing of milk, apple juice, liquid whole eggs, baking purpose and processing of green pea soup to improve the shelf life of orange juice. To enhance extraction yield of juices and bioactives from plant sources. (Canovas *et al.*, 2011).

Moreover, applied to enhance extraction of sugars and other cellular content from plant cells, such as sugar beets.

2.2 HIGH PRESSURE PROCESSING (HPP)

High pressure processing (HPP) is also referred to as Ultra-high pressure (UHP) or High hydrostatic pressure processing (HHPP). This technology has been accepted as an attractive substitute to traditional preservation methods using preservatives or thermal processing (Hernando *et al.*, 2008).

In this technology, Pressure up to 900MPa is mostly applied to kill many of the microrganisms present in foods, still at room temperature without degradation of vitamins, flavor and colour molecules in the whole practice. High pressure up to 1000MPa is applied to food packages which are immersed in a liquid, where the pressure distributed instantly and uniformly throughout the food (isostatic). In addition to, pressure up to 350MPa applied for 30 minutes or 400MPa for 5minutes caused a tenfold reduction in vegetative cells of bacteria, yeasts or moulds. (Hoover *et al.*, 1989). Usage of HPP in food products has been gaining interest as it allows inactivation of vegetative organisms at low or moderate temperature along with least degradation (Ghani and Farid, 2007).

Principle:- The basic principle of HPP is based on high pressure applied in an "isostatic" manner so that all regions of food experienced uniform pressure, unlike heat processing where temperature gradients are established.

High Pressure Processing Equipment

Main components of a high pressure system are a pressure vessel and its closure, a pressure generation system, a temperature control device and a materials handling system. Food product (either solid or liquid) packages are loaded on top of the vessel and top is closed. The pressure medium, usually water, is pumped into the vessel. As soon as, the desired pressure is attained, the pumping is stopped than the pumping is stopped, valves are closed and pressure is maintained deprived of further requirement for energy input. (Mathavi *et al.*, 2013).

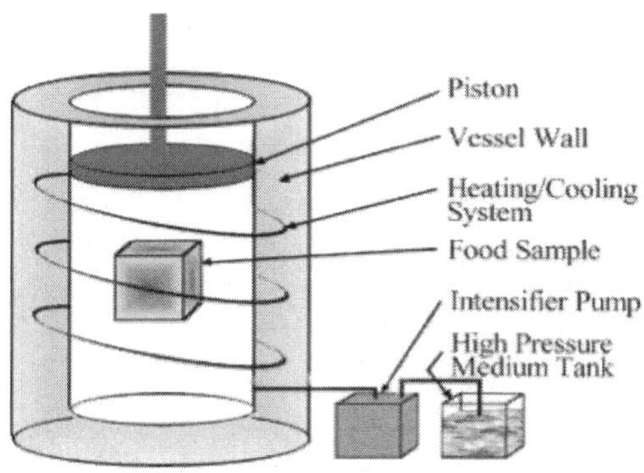

Fig.2. High Pressure Processing Equipment

Temperatures in the range from 90 to 110 C in conjunction with pressures of 500–700 MPa have been applied to inactivate spore-forming bacteria such as Clostridium botulinum. It has been reported that in HPP, pressure up to 50 MPa at 25 C for 3 days have the potential to accelerate ripening of Cheddar cheese. Likewise, these conditions were also applied to commercial Cheddar cheese. (Mathavi *et al.*, 2013). Furthermore, requirement of Pressures up to > 400 MPa cause greater than 50% denaturation of the whey protein. (Beresford. *et al.*, 1998).

Applications

HPP has been used for pasteurization and sterilization of fruits and fruit products, sauces, yoghurt, pasteurization of meat and vegetables, decontamination of high risk products, and sterilization of heat sensitive products like shellfish, flavourings and vitamins. HPP can be used not only for preservation, but also for modifying the physical and functional properties of some foods (Canovas *et al.*, 2011).

2.2 OHMIC HEATING (OH)

Ohmic Heating (OH) can be defined as a process wherein electrical current passed through a food product having an electrical resistance. Also named as resistance heating, electro-conductive heating or Joule heating (Fellows, 2009) The main advantages of ohmic heating consists of: better food product quality, required less cooking time, low capital cost, better energy efficiency. Also, an environment friendly process (Varghese *et al.*, 2014).

Many Foods or food products that contains water and ionic salts have capability of conducting electricity however, they too have a resistance that generates heat whenever an electric current passed through them. Consequently, Ohmic heating (OH) is an innovative thermal processing technique where in the food material works as an electrical resistor, moreover, heated by passing electricity throughout it (Fig. 3). Later, heated food products cooled in continuous heat exchangers, aseptically filled into pasteurized containers likewise in conventional aseptic packaging. Processing of both high and low acid food products done by this technique. Depending upon the food product's electrical conductivity whether the food product act as an insulator or a conductor, so, ohmic heating cannot be used for all the food products.

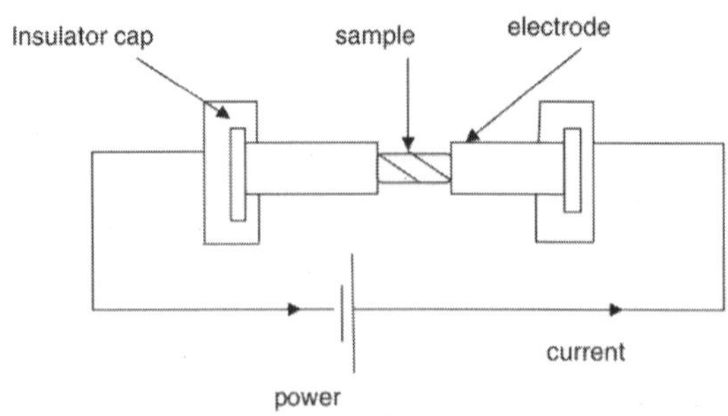

Fig . 3 Ohmic heating process.

Ohmic heating(OH) is different from other electrical heating techniques by means of the presence of electrodes contacting the food, frequency of the electric current or the waveform. The main aim behind the development of ohmic heating process was to allow HTST sterilization of solid –liquid mixtures (Canovas *et al.*, 2011). There are certain factors affecting ohmic heating such as 1) electrical conductivity of the food/ food product, 2) temperature dependence of electrical conductivity, 3) the design of heating device, 4) the residence time, 5) distribution time, 6) thermo- physical properties of foods and 7) electric field strength. (Mathavi, *et al.*, 2013).

Principle:- The principal mechanisms of microbial inactivation in ohmic heating are thermal in nature. But some evidence exists for non-thermal effects of ohmic heating as well. Microbial inactivation curves of ohmic heating is same as in conventional heating curves except for a difference in the slope likely to be explained in the presence of the electric field. Some literatures indicated that a mild electroporation mechanism may perhaps occur during ohmic heating (Lebovka *et al.*, 2005 ; Kulshrestha and Sastry 2006). Wherein low frequencies (50-60 Hz) and high field strengths (>100V/cm), the naturally porous cell walls permit the cell membrane to build up charges and formed disruptive pores. Besides, excessive exposure cause cell death owing to the leakage of intracellular components through the pores.

Applications

Ohmic heating has many applications in several food processes like cooking, blanching, pasteurization, sterilization, and extraction, specifically in high viscous or particulate foods Especially of ready-to-serve meals, fruits, vegetables, meat, poultry or fish. (Zhang , 2009). The major applications of ohmic heating used for sterilization of high proteinaceous materials, sea food processing, pre-treatments for water removal, thawing, detection of starch gelatinization and extraction enhancement. (Varghese *et al.*, 2014).

2.4 ULTRASONIC PROCESSING

Ultrasonic processing is considered as another innovative technique that has been studied for many different purposes over the last years. It has found great attention among food technologists and food research meant for its ability to affect the physical properties of food media. (Barba *et al.*, 2018). The utilization of ultrasound in food industry comprises of various novel ideas and methods which are interesting when compared to conventional methods or techniques. Their use turns into an active subject within the food industry for both of the research and development. (Povey and Mason, 1998). The basic principle behind working of ultrasound is the utilization of interaction between high-frequency sound waves and matter to find out information regarding the composition, structure and dimensions of materials through which it propagated. (Rana *et al.*, 2017)

Ultrasound can be categorized into three frequency ranges; power ultrasound (16–100 kHz), high frequency ultrasound (100 kHz–1 MHz) and diagnostic ultrasound (1–10 MHz). Application of power ultrasound marked its importance in ice cream as it aids in reduction of ice crystal size and prevention in crustation on freezing surface (Zheng and Sun, 2006). High frequency ultrasound are disruptive in nature and had greater influence on the mechanical, physical or biochemical properties of foods. Mostly applied for controlling microstructures and modification, emulsification and defoaming of food products. (Rana *et al.*, 2017). One among the emerging technology, ultrasonic processing was developed to reduce the cost and time of processing. Furthermore, ensure the quality and safety of food products. (Mason *et al.*, 2011).

Power ultrasound generation and equipment: It is an electricity driven ultrasonic system commonly employed in the food processing industry. The basic components include 1) power generator, 2) the transducers (that converts the electrical power to mechanical vibrations). There are two major types of ultrasound transducers are magneto-strictive transducers and piezoelectric transducers.

1. Magnetostrictive transducer provides a large driving force as the system is of an extremely robust construction. However, approximately 40 % of electrical energy losses occurs as heat and consequently external cooling be there required. Its maximum operating frequency is restricted to 100kHz.
2. Piezoelectric transducer has higher electro mechanical conversion and approximately 95% electrically efficient. Moreover, can be operated over the whole ultrasonic range (Rana *et al.*, 2017).

Process:- In this process, the power generator transforms electrical energy into high - frequency alternating current whereas a transducer that transforms the alternating current into mechanical vibrations acting on the surface of food specimens or travel through a liquid food media. (Barba *et al.*, 2018). Also, a delivery probe which conveys the sonic vibrations into a medium to couple sonic vibrations to the treated food material. Here, transducers may possibly take the shape of a rod, plate, bar, or sphere, commonly made from titanium, aluminium or steel. These ultrasonic transducers can be mounted external on the wall of a vessel or flow cell and stayed in indirect contact with foods else, it can be inserted into a treatment chamber or flow cell of specified geometry to transmit energy directly into a food system with better energy efficiency (Feng and Yang 2005).

When ultrasound waves passed through a liquid food media, the interaction between the ultrasonic waves, liquid and dissolved gas results into an exciting phenomenon named as acoustic cavitation (AC). The main mechanism responsible for the ultrasonic deactivation effect is the physical forces that generated via the acoustic cavitation. Later, the asymmetric collapse of a cavitation bubble leads to a liquid jet rushed through the centre of the collapsing bubble. The speed of this microjet be there in few hundred metres per second. Due to this high speed jet, pitting on solid surfaces has been observed. The Microorganisms may perhaps have hydrophobic surfaces that promote the collapse of the cavitation bubbles on the surface leading to severe damage to the cell wall. Likewise, the microstreaming leads into the erosion of cell walls that resulted in the inactivation of the microorganisms. (Chandrapala *et al.*, 2012).

Applications

Ultrasonic processing technology has a wide range of applications in the food industries, including inactivation of micro-organisms and enzymes, for pasteurization at mild heat and extraction techniques in solid materials. It is used in the improvement in efficiency of filtration and enhanced the extraction of valuable compounds from vegetables and food products also. Moreover, it has been investigated that Ultrasound-assisted process of meat tumbling had significant improvement of the yield, tenderness and juiciness of the end product. (Dolatowski and Stasiak , 1995).

2.5 MICROWAVE HEATING OR PROCESSING

Microwaves are electromagnetic waves having an electric field and a magnetic field component with frequency range from around 0.3 GHz to 300 GHz with corresponding wavelengths ranging from 1m to 1mm. Microwaves fall between radio waves and infrared radiations. For many food applications the approved and most commonly used microwave frequencies are 2450 MHz and 915 MHz. (Canovas *et al.*, 2018). In Domestic microwave ovens, 2450 MHz used. On the other hand, in industrial ovens both 2450 and 915 MHz frequencies are used.

Microwave heating is the process in which electromagnetic waves of certain frequencies used to generate heat in material. (Metaxas, 1996). When a microwavable container along with food is kept in a microwave oven and then an oven is activated. Later, the food at the edges of the container heats faster and a temperature gradient develops between the centre and the edges. Microwave energy has been applied in pasteurization and sterilization as well as in drying, thawing, tempering, blanching, and even combined to freezing (Eliasson *et al.*, 2015)

Mechanism of Microwave heating

Heating with microwaves primarily involves two mechanisms:

1. Ionic polarization:

When an electrical field is applied to food solutions containing ions (i.e. dissolved salts), these ions move at an accelerated pace due to their inherent charge. Due to which collisions between the ions occurs that generates heat. (Ohlsson, 1993)

2. Dipole rotation

Food materials contain polar molecules such as water. Due to their dipolar nature, water molecules generally have a random orientation. Once microwave energy/ electric field is applied, water molecules inside the food orient themselves according to the polarity of the field. Such rotation of water molecules resulted in friction with the surrounding medium, thus, heat is generated. (Ohlsson, 1993)

Microwave Oven

Microwave oven consists of the following components:

Power supply and control:- the main purpose of the power supply is to draw electrical power from the line. Also, converted in to the high voltage required by the magnetron. It controls the power to be fed to the magnetron as well as the cooking time

Magnetron or power tube:- Magnetron is an oscillator capable of converting the power supplied into microwave energy. It is called the heart of the microwave oven. Frequency of 2450 MHz has been set aside for microwave oven for home use.

Waveguide or transmission section:- It is a rectangular metal tube which directs the microwaves generated from the magnetron to the oven cavity. It helps to prevent direct exposure of the magnetron to any spattered food which would interfere with function of the magnetron. The wave guide is a few centimetres long in a domestic oven whereas it can be a few meters long in industrial units. Usually, the energy loss in the waveguide is quite small.

Stirrer:- It is usually a fan-shaped distributor commonly used to distribute microwaves from the waveguide and allow more uniform heating of food. In general, the stirrer disturbs the standing wave patterns, thus allowing better energy distribution in the oven cavity.

Turntable:- It rotates the food products through the fixed hot and cold spots inside the cooking cavity and allows the food products to be evenly exposed to microwaves

Oven cavity:- The oven cavity is a space that encloses the food to be heated within the metallic walls once exposed to microwaves.

Door and choke:- It allows the access of food to the cooking cavity. The door and choke are specially engineered which prevent microwaves leakage through the gap amid the door and the cooking cavity.

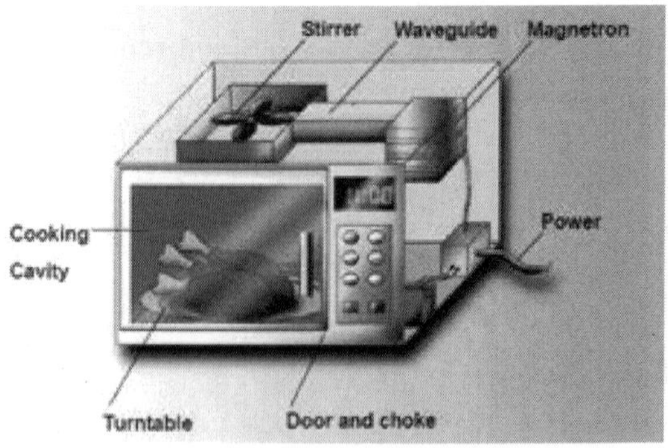

Fig. 4. Components of microwave-oven.

Applications

Microwave is used in baking, concentration, cooking, curing, drying, finish drying, freeze drying, pasteurising, sterilizing, tempering and thawing.

2.6 Radio Frequency

The Radio- frequency (RF) radiation is lying at a higher frequency than MWs in the spectrum of electromagnetic fields. When RF is applied to foods, different types of interactions both intrinsically as well as between the applied electromagnetic field and the constituent charged moieties occur within the material. These interactions are described as electronic, atomic, ionic, and molecular displacements. More specifically, the move of the electrons with respect to the positive nucleus of each atom can be defined as electronic displacement. Atomic displacement involves interactions between the atoms of each molecule or just between the atoms in case of a material with crystalline structure (Xanthakis & Valdramidis, 2017).

The two main mechanisms occurring during RF heating are the ionic and the molecular displacement. When an electric field is applied to a dielectric material, the positive ions are forced to move to one direction, while the negatively charged ions to the opposite. Therefore in case of an alternate electric field the ions tend to move forward and backward led by the enduring antistrophe of the polarity according to the frequency of the applied field. This internal turmoil driven by the indecisive external factor leads to the generation of heat within the product by friction. The aforementioned phenomenon is described as heat generated by ionic displacement. Molecular displacement involves molecules with a native asymmetrical form and unbalanced charge distribution. These molecules are characterized by electric moment in addition to their autogenous electric field when an external electric field is applied. The molecules are polarized and attempt to realign with respect to the electric field vector. The oscillation of the polar molecules when an alternate electric field is applied produces heat within the material due to friction forces.

The inactivation mechanism of food-borne pathogens by RF treatment is mainly caused by heat generation within the food. Up to now, RF processing has been proposed for pasteurization of dairy and pasta products, for the sterilization of low acid foods, microbial inactivation of salmon caviar, ground beef pathogens, and other food materials (Al-Holy, Wang, Tang, & Rasco, 2005; Schlisselberg et al., 2013; Wang et al., 2003).

2.7 INFRA-RED HEATING

Infrared is a kind of electromagnetic radiation that lies between ultraviolet and microwave energy region. Based upon its spectral range, infrared radiations are normally categorized into near-infrared (700-1400 nm), mid-infrared (1400-3000 nm), and far-infrared (3000-10000 nm) regions (Maloney and Harrison 2016, Rastogi 2015). Far-infrared is the most suitable for food processing because most food constituents absorb radiation in the far-infrared region (Rastogi 2012, Wang et al. 2014). Infrared (or radiant) heating is an indirect mode of heating wherein electromagnetic energy penetrates the food, gets adsorbed on the surface and then converts to heat. The heat adsorbed on the food surface is mostly by radiation but to a lesser extent by convection and conduction mechanism. The magnitude of heating by radiant energy depends upon the food surface characteristics as well as food color, therefore, IR radiation is typically used to alter the food quality by modifying the flavor, aroma and surface color of the food products. IR rapidly and uniformly heats the product which not only reduces the processing time and energy costs but also prevents the product overheating because of rapid heating rates. The temperature of the air inside the instrument can be kept constant because the air is not heated by IR which helps to controls the product overheating during processing (Wang et al. 2014, Mao et al. 2011, Maloney and Harrison 2016).

Due to its compact design with high controllability and safety, IR heating has been widely adopted in the food industry for cooking, frying, drying, dehydration, roasting, baking, peeling, blanching, and pasteurization of agricultural and food products (Rastogi 2012, Moreirinha et al. 2016, Ramaswamy, Krishnamurthy, and Jun 2012).

The potential of this technology has only been exploited to a limited extent for heating purposes in the food industry. The technology can penetrate and supply heat to only a few millimeters below the surface of a sample which limits its application for heating a small number of food product (Rastogi 2015, Rastogi 2012). Additionally, this poor penetration capacity of IR slows down the temperature increase of solid foods as their thermal conductivity (k) is much lower than the liquid foods. To make the penetrative radiation energy more effective, IR heating may be used in combination with other conventional modes of heating for applications such as freeze drying, dehydration, cooking and baking (Wang et al. 2014, Mao et al. 2011).

2.8 COLD PLASMA TREATMENT

Cold plasma technology (CPT) is a novel and emerging non-thermal processing technology that uses energetic, reactive gases to inactivate pathogenic and spoilage microorganisms pertinent to food. The technology was initially developed to enhance the surface energy of polymer and sterilization of medical equipment in hospitals (Pankaj et al. 2014, Bahrami et al. 2016, Jayasena et al. 2015). However, it has recently emerged as a powerful disinfection tool for food industry for in-package and post-packaging decontamination of food products including the dry disinfection of solid and liquid food surfaces like dried milk, meat, poultry, fish, herbs, sprouted seeds, grains, spices and fresh produces (Jayasena et al. 2015, Korachi et al. 2015, Misra et al. 2011, Lee et al. 2015, Scholtz et al. 2015).

While cold plasma technology is gradually gaining acceptance among food processors, the long lasting effect of generated reactive species and their actual mechanism is still unclear. In some cases, reactive species change the morphology of biological cells and cause hindrance in their regular functions (Jayasena et al. 2015), and the role of these active species on some sensitive food constituents such as lipids and vitamins is still ambiguous (Scholtz et al. 2015). Some of the reactive species trigger the oxidation of high lipid containing food products which produce off flavor compounds that cause rancidity. Therefore, meat products are not considered an ideal substrate for plasma treatment (Awad et al. 2012). Nevertheless, cold plasma treatment is an emerging food processing technology which is rapid and does not leave any toxic residuals or exhaust gases post-processing. However, issues regarding the nutritional content, color, texture, chemical changes and overall food quality need to be considered (Mason, Chemat, and Ashok Kumar 2015, Korachi et al. 2015). Although, CPT is not fully adopted by the food industry for large scale industrial setting due to the lack of knowledge on some critical parameters, the equipment is readily scalable and has potential for wide-scale applications. Research efforts around the globe are underway to understand the safety of the gases used before bringing it for commercial usage (Awad et al. 2012).

Conclusion

As we know the main problem with the thermal processing of food is loss of volatile compounds, nutrients, and flavor. To overcome these problems novel technologies came into food industries to increase the production rate and profit. The non thermal processing is used for all foods for its better quality, acceptance and for its shelf life. These emerging technologies mainly rely on physical processes that use ambient or moderately elevated temperatures and short treatment times to inactivate microorganisms and are mostly employed to the liquid packed foods as compared to solid foods. The cost of equipments used in novel processing is high compared to equipments used in thermal processing. After minimizing the investment costs of non thermal processing methods, it can also be employed in small scale industries. More research is required to establish and expand the industrial applications of the different technologies and overcome high capital costs that may delay industrial adoption.

REFERENCES

- Barba, F. J., Ahrne, L ., Xanthakis, E. , Landerslev, M.G and Orlien, V. 2018. Innovative Technologies for Food Preservation. Elsevier:25-51.
- Barba, F. J., Parniakov, O., Pereira, S. A., Wiktor, A., Grimi, N., Boussetta, N., and Vorobiev, E. 2015. Current applications and new opportunities for the use of pulsed electric fields in food science and industry. *Food Research International*, **77** (4): 773- 798.
- Beresford, T. P., O. Reilly, C., O.Connor, P., Murphy, P. M and Kelly, A. 1998. Acceleration of cheese ripening: current technologies, potential use of high pressure. *Proceedings of VTT Symposium Helsinki, Finland*: 186, 103 - 114.

- Canovas , G. V. B., Albaali, A. G., Juliano, P and Knoerzer, K. 2011. Introduction to Innovative Food Processing Technologies: Background, Advantages, Issues, and Need for Multiphysics Modeling. Chapter. *Innovative Food Processing Technologies: Advances in Multiphysics Simulation,* First Edition. Edited by Kai Knoerzer, Pablo Juliano, Peter Roupas, and Cornelis Versteeg. *John Wiley & Sons, Ltd*: 3-21.
- Chandrapalaa, J., Oliverb, C., Kentishc, S and Ashokkumara, M. 2012. Ultrasonics in food processing, Food quality assurance and food safety. *Trends in Food Science & Technology*. **26** :88-98.
- Dolatowski ,Z. J. and Stasiak D. M, 1995. Tumbling machine with ultrasound. *In: The 9th Congress of Food Science and Technology*, Budapest, 153.
- Eliasson, L., Isaksson, S., Lovenklev, M and Ahrne, L. 2015. A comparative study of infrared and microwave heating for microbial decontamination of paprika powder. *Frontiers in Microbiology*, 6.
- Fellows, P. J. 2009. Food processing technology: Principles and practice (23rd edn). Cambridge: Woodhead Publishing Limited and CRC Press LCC.
- Feng, H and Yang, W .2005. Power ultrasound. In: YH Hui ed., Handbook of Food Science, Technology and Engineering, New York : CRC Press: 121 – 9.
- Ghani, A. G and Farid, M. M. 2007. Numerical modeling of heat transfer in high pressure processing using computational fluid dynamics. *Proceedings of the 4th International Conference on High Pressure Bioscience and Biotechnology*. 1: 205 – 212.
- Hernando, S. A , Tarrago, M. S , Purroy, B. F and Tonello, S. C. 2008 . Advances in design for successful commercial high pressure food processing. *Food Austrailia*, **60** (4): 154 – 156.
- Hoover, D. G., Merick, C., Papineau, A. M., Farkas, D. F and Knor, D. 1989. Application of high hydrostatic pressure on foods to inactivate pathogenic and spoilage organisms for extension of shelf life. *Food Technology*, **43**(3): 99.
- Knorr, D., Froehling, A., Jaeger, H., Reineke, K., Schlueter, O and Schoessler, K. (2011). Emerging technologies in food processing. *Annual Review of Food Science and Technology*, **2**: 203-235.
- Knoerzer K. Food Process Engineering. Reference Module in Food Science. Elsevier, 2016. https://doi.org/10.1016/B978-0-08-100596-5.03333-3
- Kulshrestha, S. A and Sastry, S. K. 2006. Low - frequency dielectric changes in cellular food material from ohmic heating: Effect of end point temperature. *Innovative Food Science and Emerging Technolologies*, **7** (4): 257 – 262 .
- Lebovka, N. I., Praporscic, I., Ghnimi, S and Vorobiev, E . 2005. Does electroporation occur during the ohmic heating of food. *Journal Food Science*, **70** (5): 308 – 311.

- Mason, T. J., Chemat, F and Vinatoru, M. 2011. The Extraction of Natural Products using Ultrasound or Microwaves. Current Organic Chemistry. **15**:237-247.
- Mathavi, V., Sujatha, G., Ramya, B and Devi, K. 2013. NEW TRENDS IN FOOD PROCESSING, *International Journal of Advances in Engineering & Technology,* **5** (2): 176-187.
- Metaxas, A. C. 1996. Foundations of Electroheat . Chicester : J. Wiley & Sons.
- Ohlsson, T. 1993. Domestic use of microwave ovens. *Encyclopaedia of food science food technology and nutrition. 2: 1232-1237.*
- Pereira, R. N and Vicente, A. A. (2010). Environmental impact of novel thermal and non-thermal technologies in food processing. Food Research International, 43(7), 1936- 1943.
- Povey, M. J. W and Mason, T. J. 1998. Ultrasound in food processing, London, Blackie Academic and Professional.
- Puertolas, E and Barba, F. J. 2016. Electrotechnologies applied to valorization of by-products from food industry: Main findings, energy and economic cost of their industrialization. *Food and Bioproducts Processing,* **100** (A): 172-184.
- Rana, A., Meena and Shweta. 2017. *Ultrasonic processing and its use in food industry: A review.* International Journal of Chemical Studies, **5** (6): 1961-1968.
- Varghese, K. S., Pandey, M. C., Radhakrishna K and Bawa, A. S. 2014. Technology, applications and modelling of ohmic heating: a review. *Journal Food Science Technology,* **51**(10):2304–2317.
- Yeom, H. W., Streaker, C. B., Zhang, Q. H. and Min, D. B. 2000. Effects of pulsed electric fields on the quality of orange juice and comparison with heat pasteurization. *Journal of Agricultural Food Chemistry,* **48**: 4597-4605.
- Zhang, H. 2009. Electrical properties of foods In G. V. Barbosa-Canovas (Ed.), *Encyclopedia of life support systems-food engineering,* **1**: 110-119.
- Zheng, L. and Sun, D. W. 2006. Innovative applications of power ultrasound during food freezing processes – a review. *Trends in Food Science Technology,* 17:16-23.
- Zimmermann, U. (1986). Electrical breakdown, electropermeabilization and electrofusion. *Reviews of Physiology Biochemistry and Pharmacology,* 105, 176-256.

CHAPTER 4

FOOD BIO-PRESERVATION

Shafaq Javid*[1], Anju Bhat[1], Entesar Hanan[2], Nadira Anjum[1], Fozia Hameed[1] and Anjum Ayoub[1]

[1]*Division of Food Science and Technology, Sher-e-Kashmir University of Agricultural Sciences and Technology- Jammu, Chatha-180009, India*

[2]*Department of Food Technology, Jamia Hamdard, Hamdard Nagar, New Delhi-110062, India*

Corresponding Author: Email ID: shahshafaq14@gmail.com

*Corresponding Author: *1s.fozia011@gmail.com*

ABSTRACT

Food spoilage is considered as a major issue for food business operators as it leads to wastage of food. It affects the brand name and at the same time results in economic losses for both manufacturers and consumers. Due to the growing consumer demand for safe, nutritious and minimally processed foods, the traditional methods of controlling the hazards in food are nowadays being replaced by newer and innovative techniques of food preservation. Food – Bio preservation is one among the innovative techniques. Bio-preservation of involves the use of certain antagonistic microorganisms or the metabolic products produced by them to retard, delay or inhibit the activity of the undesirable microorganisms in food products to enhance the shelf life of the products and to maintain the quality as well as the safety of food. In the field of bio-preservation, bacteriocins particularly produced by lactic acid bacteria have received much attention due to their ability to be used as natural preservatives in food industry. This technique can be used in combination with other preservative factors in an effective way to achieve food safety and quality.

Keywords: Bio-preservation, food, bacteriocins, spoilage, quality

1. INTRODUCTION

Food safety standards in food processing and microbiological have condensed but not eliminated the spoilage of product and food related illness in industrialized countries. Spoilage of food causes loss in texture, flavor and damage to the nutritional value of food that ultimately becomes unsuitable to eat as well as harmful to people. One of the major concerns in food processing industries is the contamination by pathogens which ultimately causes food born illness. With growing demand for pre cooked foods like seafood, gets damaged due to temperature fluctuations. On the other hand export of raw sea food in developing countries results in outbreak of food borne illness (Nath *et al.*, 2013). More than 250 to 350 million people were affected in USA by gastroenteritis, near about 500 human deaths, 20 % to 30% are considered due to food borne diseases with the main foods like meat, fish , poultry, sea food, dairy products(Mead *et al.*,1999). Various food industries makes use to chemical preservatives and physical treatments (like high temperature) in order to attain food safety standards against bacterial pathogens. But on the other hand these preservatives have many draw backs also which includes modification and alteration in organolaptic and nutritional quality of food, toxicity of chemical preservatives, and growing demand of customers for safe and minimally processed food without using additives. In order to control microbial spoilage and to bring necessary food safety standards various combination of innovative technologies have been introduced that include biological antimicrobial systems such as lactic acid bacteria (LAB) and/or their metabolites to synchronize consumer demands with the necessary safety standards that include biological antimicrobial systems such as lactic acid bacteria (LAB) and/or their metabolites Now a day's Food safety can be achieved by replacing chemical additives by natural products without harming the host or the environment. Use of biotechnology in food processing sector targets the assortment, construction and enhancement of useful microorganisms and their products, making use of non pathogenic micro organisms and their metabolites in order to extend the shelf life of food including their technical application in food quality (De Martinis and Franco, 1998). Antagonistic properties of LAB allied to their safe history of use in traditional fermented food products make them very attractive to be used as biopreservatives. On the other hand food borne illness caused due to fungi are the major issues in food processing sector due to their ability of growing at any stage or even during harsh environment(Caplice and Fitzgerald ,1999).Some fungal genera such as *Alternaria, Aspergillus, Penicillium* and *Fusarium* can produce secondary metabolites that have toxic effect on humans like mycotoxins, which can resists various food processing steps and thus leads to food safety concerns and can cause various negative effect on quality of food products. There are 3 main stages where fungal contamination can asses (i) the field, where water, soil, and air are natural fungal niches; (ii) fungal growth during transportation, handling ,storage even during packaging in various food materials like meat, fish, poultry during manufacturing and processing of bakery and dairy products(Pitt and Hocking, 2009) . Synthetic fungicides along with crop rotation practices can be used in order to control fungal contamination. Various modern techniques used in packing of food are also used to protect crops from mechanical damage where microbes can attack very easily. (Rojas-Graü *et al.*,2009). Many chemical preservatives like propionate, benzoate, sorbate, nitrate, nitrite and sulfites are used to avoid fungal spoilage (Silva and Lidon, 2016). The most commonly used antifungal compound used as food preservative is natamycin (E235). It is produced by Streptomyces. Adverse health and environmental effects due to such fungicides and preservatives have led to develop more natural products. Number of products is commercialized in marked composed of bacteria, fungi and yeast (Ness and Holo, 2000). Due to increased demand for more natural foods and less processed products, more emphasis are laid on consumer safety, good hygienic practices and quality of food product by replacing chemical preservatives. All the safety measures like safety estimation, organolaptic objectivity, and activity stability of the bioprotective cultures needed to be evaluated before rising in market. Studies are still in progress to know the overall action and the whole mechanism of antifungal compounds and their effect on human health.

2. BIO- PRESERVATION

In order to increase microbiological safety of food non pathogenic micro organisms and their metabolites are used to extend the shelf life of product, known as bio preservation. Bio-preservation can be applied in order to increase the shelf life, extend the storage and enhance the safety of foods by using natural micro flora and (or) their antibacterial products. Frementation is one of the most common form of bio preservation in which there is break down of complex compounds which causes production of various compounds like acids, alcohol, synthesis of Vitamin-B12, riboflavin and Vitamin-C precursor which ensures antifungal activity and improvement of organolaptic qualities such as, production of flavor and aroma compounds whether fermentation is added or natural. Bio preservation in case of fish processing can be achieved by using adding microbial or by increasing acidity of fish muscles. Defensive type of bacterial cultures have been determined and developed along with anti microbial agents against various pathogens and spoilage micro organisms. Various compounds contributing to antimicrobial activity diacetyl, organic acids, bacteriocins and acetaldehyde, enzymes, CO_2, hydrogen peroxide etc by microbiota.

Table 1: Some natural antimicrobials for food bio-preservation

Antimicrobials	Sources	Food Biopreservation
Organic acids	End product of fermentation	Decrease the pH of surrounding environment, creating a selective barrier against nonacidophiles
Co_2	Buy product of sugar fermentation by heterofermentative LAB	Creation of anaerobic bacteria against aerobic bacteria and produces carbonic acids.
Hydrogen peroxide	Produces LAB in presence of oxygen	Anti bacterial effect through oxidative damage of protein and increase membrane permeability.
Reuterin	Antimicrobial compound produced by*Lactobacillus reuteri*	Antimicrobial activity against yeast, bacteria and mold by inhabiting DNA synthesis

Source: Silva and Lidon, 2016

The various biopreservatives are:

2.1 BACTERIOCINS

In 1925 Andre Gratia discovered first bacteriocin as Colicine. As bacteriocin are proteinaceous agents which are digested by proteases in human digestive tract. They are ribosomally synthesized; their characteristics can be improved in oder to enhance their intensity and mode of action. As they are peptides or complex proteins and are produced by bacteria, not generally termed as antibiotics which have therapeutic or medicinal use (Nath *et al.*, 2013). Bacteriocin can lead to benefits when they are in suffient amounts in many foods and feed producers, these peptides can kill or inhibit pathogenic bacteria that compete for the same ecological niche or nutrient pool. As many bacteriocin have narrow range and thus effective against related bacteria with nutritive demands for the same scarce resources. The important category of bacteriocins is LAB bateriocins. For mutation and selection of bacteriocin variants Recombinant DNA technology is currently applied to enhance production and to transfer bacteriocin genes to other species with increased /or broad activity spectra. In many cases where potential of bacteriocin during food preservation are exploited, their physical and chemical properties, mode of action and structure-function relationships of bacteriocin should be considered thoroughly in order to avoid various human risks. Synergistic reactions of these compounds and other natural preservatives along with advanced technologies will help in the replacement of chemical preservatives or could help in minimal heat treatments during processing, meanwhile maintaining adequate microbiological safety and quality in foods (Piard, et al., 1992).

Table 2: Common bacteriocins and their possible uses in food industry.

Bacteriocins	Food applications
Nisins	*Clostridium butulinum* in cheese, *L. monocytogenes* in cheeses such as Camembert, Ricotta and Manchego
Pediocin	Effective against milk and cheddar cheese *L.monocytogens.*
Pediocin PA-1/AcH	In salad and fruits juices
Nisins	In vegetable and fruit juices.

Source: Cleveland, et al., 2001

2.1.1 Classification of bacteriocins

Bacteriocin is classified into 4 major groups on the basis of their action and structure .

2.1.11 Class I

- Lantibiotics which consists of small group of peptides and are characterized by several unusual amino acids (Guder ,*et al.*, 2000).
- Bacteriocins are small peptides by their content in dehydro amino acids and ether amino acid as compared to other bacteriocins.Which include nisin, lacticin 481 of *L. lactis* , lacticin 3147 of *L. lactis* , *citolysin* of *E. faecalis* .

2.1.1.2 Class II

- These bacteriocins are heat stable peptides small, non modified, heat stable peptides.
- Due to amphiphilic helical structure, they can insert into the membrane of target cell which ultimately leads to depolarization and death.
- They comprises the (<10 kDa) thermostable non-lantibiotic linear peptides.

2.1.1.3 Class III

- They are termed as bacteriolysins, large (> 30 KDa), heat-labile protein bacteriocins such as helveticin J of L. helveticus and bacteriocin Bc-48 of E. faecalis.
- They can directly act on cell wall of gram positive bacteria thus leading to death and lysis of the target cell.

2.1.1.4 Class IV

- Class 4 is reserved for cyclic bacteriocins composed not only from protein (also lipid or cidrate).

2.1.2 Factors inhibiting bacteriocin production include:

- Adverse environmental, physical conditions and chemical composition of food (like pH, temperature, nutrients, etc.);
- Impulsive loss in production capacity.
- Inactivation by phage of the producing strain
- Antagonism effect of other microorganisms in foods.

2.1.3 Features of bacteriocin considered for food applications

Various features are considered while selecting bacteriocin-producing strains for food applications:

- Strain used should GRAS (generally regarded as safe) certified.
- Broad spectrum of inhibition that includes pathogens or else high specific activity depending upon application.
- Should have thermo stability.
- Should have beneficial effects and improved safety.
- No adverse effect on quality and flavor.

2.1.4 LAB bacteriocins

Bacteriocin belonging to the class 11 are small (about 30- 100 amino acids) are heat stable but commonly not post translationally modified like (plantaricins of *Lactobacillus plantarum* C11 and *sakacin P* of *Lactobacillus sake*) can be transcriptionally regulated through a signal transduction system, as it consists of three components an induction factor (IF), histidine protein kinase (HK) and a response regulator (RR). Some bacteriocin can sythezise on bacteria while LAB produces multiple bacteriocin bacteriocins (2–3 bacteriocins). A large number of new bacteriocins in lactic acid bacteria (LAB) have been characterized in recent years. Many bacteriocin producing strains can be used as protective cultures in variety of food products. They posses many characteristic features that make them suitable candidates for use as food preservatives. Like

- Nature of protein, inactivation by proteolytic enzymes of gastrointestinal tract.
- Non-toxic to laboratory animals tested and generally non-immunogenic.
- Inactive against eukaryotic cells.
- Generally thermo-resistant (can maintain antimicrobial activity after pasteurization and sterilization).
- Broad bactericidal activity affecting most of the Gram-positive bacteria and some, damaged, Gram-negative bacteria including various pathogens such as *Listeria monocytogenes, Bacillus cereus, Staphylococcus aureus* and *Salmonella.*

2.2 BACILLUS AS BIOPRESERVATIVE

Production, isolation and characterization from bacteriocin include *Subtilin* from *B. subtilis, Megacin* from *B. megaterium and Thermacin from B. stearothermophilus.* Bacteriocins produced by *Bacillus* spp. could be alternatives to those produced by lactic acid bacteria for several reasons:

- One of most common Bacillus *spp*, like lactic acid bacteria, strains of LAB have been used over hundred years in making food and various other enzymes intensively used in food processing. They do not produce any type of adverse effect on humans and hence are safe to use with many health benefits.
- Bacillus *spp.* has broader spectra then LAB and its bacteriocins as they act as strong anti microbial agent against Gram positive and Gram negative bacteria.
- Inhibitory activity at alkaline, acidic level in bacteriocins can be caused due to metabolic diversities after thermal processing and would be suitable for food processing.
- The physiology/genetics of Bacillus are well understood, second only to those of Escherichia coli. Molecular biological techniques would provide safe/reliable tools for producing bacteriocins for the food industry.

2.3 LACTIC ACID BACTERIA

LAB belongs to heterogeneous group of Gram-positive, low-GC, acid-tolerant bacteria. During fermentation they produce lactic acid which is major metabolic end product. They are found in various biostopes like as environment, plants, human, and animal microbiota. LAB belongs to Lactobacillales order which include 6 families, 36 genera, and more than 200 species. They are generally associated in making many food products that help in increasing shelf life, having many health benefits, organolaptic properties and nutritional parameters. The main LAB in fermented foods correspond to species belonging to the Lactobacillus, *Oenococcus, Lactococcus, Leuconostoc, Carnobacterium, Enterococcus, Pediococcus, Streptococcus, Tetragenococcus, Vagococcus, and Weissella* genera.

2.4 PROPIONI BACTERIA

Propionic bacteria are anti fungal agents belonging to the Actinobacteria class is a gram positive bacteria, high GC content divided into 'cutaneous' and 'classical' (also referred to as 'dairy') propioni bacteria based on their main isolation biotopes (Stackebrandt *et al.*, 2006). Propionibacterium genus consisting of *P. freudenreichii*, the main species used in cheese-making, and three other species, and a novel genus—*Acidipropionibacterium*—which encompasses the former species *P. jensenii, P. acidipropionici, P. thoenii,* and three other species (Scholz, and Kilian, 2016).

2.5 BACILLUS AND OTHER BACTERIA

Bacilli are found in diverse environments, gram-positive, aerobic endospore-forming and rod-shaped bacteria produce a large no of secondary metabolites which includes anti microbial agents which are known for their medicinal effects. It also acts as biocontrol agents on various pre harvests and post harvest disease related to many fruits and vegetables. Strains of Bacillussubtillus produce many anti microbial agents like peptides and non peptides. Bacillus *sp.* strains present a large activity spectrum against pathogenic fruit molds from the *Colletotrichum, Aspergillus, Alternaria, Beltraniella, Botryosphaeria, Botrytis, Monilinia, Fusarium, Fusicoccum, Penicillium, Phomopsis, Phoma, Rhizoctonia, and Rhizopus genera.*

2.6 BACTERIOPHAGES

Bacteriophages or phages are viruses that exclusively infect and multiply in bacteria, harmful to human's animals, and plants. They are most abundantly found microorganisms on earth and wide spreaded on foods of various origins (Carlton, *et al.*, 2005). Most of the bacteriophages are belong to the Caudovirales order and conatins icosahedra head and a double-stranded 179 DNA tail. Bacteriophages are the most promising anti microbial agents and have potential for reduction of pathogens in live stock farming, during milking and after slaughtering. Bacteriophage fulfill various parameters like, preventing or inhabiting of colonization, control of diseases in livestock through phage therapy, to disinfect equipment and contact surfaces by phage bio-sanitation and enhance the shelf life of manufactured foods by phage biocontrol and decontamination of carcasses and other raw products such as fresh fruit and vegetables. Bacteriophages are effective against *Salmonella* and *Campylobacter* in chicken and pathogenic *E. coli*) in ruminants. To achieve biosanitation reduction in biofilm can be obtained (Azeredo and Sutherland, 2008) by reducing or eradication of *S. aureus* nasal or skin colonisation in food handlers (Atterbury, *et al.*, 2005).

In biopreservation, bacteriophages act as an effective method in preparation of food processing's as they work symbiotically with fermenting organisms during fermentation. and hence reduce the effect of *Salmonella* in cheddar production, *S. aureus* growth in curd manufacturing and complete abolition of *Listeria monocytogenes* during ripening, and storage of acid coagulated and semi-hard cheeses and inhibition of *Enterobacter sakazakii* in reconstituted infant formula milk and Salmonella typhimurium on chicken frankfurters.

2.7 ENDOLYSINS

Endolysins are bacteriophage (phage)-encoded peptidoglycan hydrolases that are synthesized at the end of the phage multiplication cycle. Bacteriophages lyse the host bacterial cell wall and release newly assembled bacteriophage virions (Loessner *et al.,* 2005).Endolysins directly target various bonds in the bacterial cell wall peptidoglycan structure. During this process holin proteins assist their entry into the cytoplasmic membrane where endolysins lyse the bacterial cell wall. Endolysins can act as exolysins in the Gram-positive bacteria peptidoglycan layer as they are not able to degrade the bacterial outer membrane around Gram-negative bacterial cells. Indeed, the outer membrane of Gram-negative bacteria effectively prevents access of the endolysin. Therefore, researchers have attempted to develop new methods for using endolysins against Gram-negative pathogens. Endolysins from Gram-positive phages have a modular structure composed of enzymatically active domains (EADs) and a cell wall-binding domain (CBD).The EADs provides the actual enzymatic activity that cleaves the peptidoglycan structure, whereas the CBD recognizes and leads the endolysin to the speci c cell wall-associated ligand molecules with high speci city. Endolysins act as a potential have biocontrol agents which helps to prevent contamination against various pathogens in food system (Chang, *et al.,* 2017). Safety and stability of endolysins is more important regarding food products and especially in food processing system. They can be easily added in food products if they are in purified form and are usually screated by *Lactococcus lactis* or *Lactobacillus spp* .

8. ANTIFUNGAL MICROBIAL AGENTS

Anti fungal microbial agents are used to control fungal spoilage which is basic tool in order to know the fungal contaminants and their occurrence in food. Isolation and identification which will also help to know their mycotoxigenic nature .In case of fungal spoilage contamination is directly linked to given or limited number of species and the culture depended methods are also very important. The very first step is the enumeration or isolation of these micro organisms not only from the food which has been contaminated by fungal spoilage but also from other potential sources like air and other surfaces (dos Santos, et al., 2016). Whereas,to know the total counts of microphages and to isolate them specific media is used. On the other hand Pitt test has become gold standard for the identification of molds of the internal transcribed spacer (ITS), PCR ampli cation and sequencing of the D1/D2 domain of the large subunit of ribosomal RNA (26S) are routinely performed for yeast and molds, respectively. In case of molds ITS,is not that informative for some species therefore other gene targets can be sequenced to complete identification at species level.

Table 3: Active bioprotective cultures against fungal contaminants of postharvest fruits and processed and raw foods

Food field	Group	Antifungal microorganism	Source
Bakery	LAB	*Lactobacillus brevis* ITM18	Sourdough
Bakery	LAB	*Lactobacillus plantarum* HD1	Kimchi
Bakery	LAB	*Lactobacillusamylovorus* DSM19280 (1/1	Cereal isolate
Bakery	LAB	*Lactobacillus amylovorus* DSM19280 (1/3	Cereal isolate
DAIRY	LAB	*Lactobacilluscasei* AST18 (1/1	Chinese dairy products
Malting	LAB	*Lactobacillus brevis* R2Δ (1/1	porcine isolate
Fermented vegetables	LAB	*Lactobacillus fermentum* YML014	gari, fermented cassava (starchy root
Grain/ seeds	LAB	*Lactococcus sp.* BSN307	Wheat grains
Fruits	LAB	*Lactobacillus paracasei* ŁOCK0921 (1 tested in situ) (1/(9 in vitro/60))	Plant and human

Source: Oliveira, et al., 2014

Conclusion

Biopreservatives offers a great potential in order to extend the storage life and food safety, microbial stability, nutritional qualities regarding food products by using their natural micro flora or their anti microbial agents. Principal hurdles applied during food safety includes temperature, redox potential, water activity, ph,HHP, UV and competitive flora (LAB producing antimicrobial compounds) chemical preservatives etc. Various parameters like safety assessment, activity stability of the bioprotective cultures, organoleptic neutrality need to be evaluated before it reaches to the final customer. Therefore bio-preservation of food by bacteriophages, bacteriocins and endolysins has promising role in food processing preservation and food safety. Genetically modified bacteriophages may also be helpful in bio-preservation but their safety issues must be addressed properly before selection as bio-preservative. In order to attain overall advantages of bio preservatives it is necessary to have the freshness from resistance particularly of bacteriocins and bacteriophages. Endolysins act as good disinfectants with proper handling strategies. Now a day's food borne illness has become a serious issue. Therefore great emphasis laid on food safety and safety standards in food processing industries. So revolution in genetic technology and advancement in molecular techniques bacteriocins, bacteriophages and endolysins will prove the future bio-preservative.

REFERENCES

- Atterbury, R.J., Dillon, E., Swift, C., Connerton, P.L., Frost, J.A., Dodd, C.E., Rees, C.E. and Connerton, I.F. 2005. Correlation of Campylobacter bacteriophage with reduced presence of hosts in broiler chicken ceca. *Applied and Environmental Microbiology*, **71**(8): 4885-4887.
- Azeredo, J. and Sutherland, I.W. 2008. The use of phages for the removal of infectious biofilms. *Current pharmaceutical biotechnology*, **9**(4): 261-266.
- Caplice, E. and Fitzgerald, G.F. 1999. Food fermentations: role of microorganisms in food production and preservation. *International journal of food microbiology*, **50**(1-2): 131-149.
- Carlton, R.M., Noordman, W.H., Biswas, B., De Meester, E.D. and Loessner, M.J. 2005. Bacteriophage P100 for control of Listeria monocytogenes in foods: genome sequence, bioinformatic analyses, oral toxicity study, and application. *Regulatory Toxicology and Pharmacology*, **43**(3): 301-312.
- Chang, Y., Kim, M. and Ryu, S. 2017. Characterization of a novel endolysin LysSA11 and its utility as a potent biocontrol agent against Staphylococcus aureus on food and utensils. *Food microbiology*, **68**: 112-120.
- Cleveland, J., Montville, T.J., Nes, I.F. and Chikindas, M.L. 2001. Bacteriocins: safe, natural antimicrobials for food preservation. *International journal of food microbiology*, **71**(1): 1-20.
- De Martinis, E.C. and Franco, B.D. 1998. Inhibition of Listeria monocytogenes in a pork product by a Lactobacillus sake strain. *International Journal of Food Microbiology*, **42**(1-2): 119-126.
- dos Santos, J.L.P., Bernardi, A.O., Morassi, L.L.P., Silva, B.S., Copetti, M.V. and Sant'Ana, A.S. 2016. Incidence, populations and diversity of fungi from raw materials, final products and air of processing environment of multigrain whole meal bread. *Food Research International*, **87**:103-108.

- Guder, A., Wiedeman,I. and Sahl, H.G.2000.Post translationally modified bacteriocins, the lantibiotics. *Bioploymers*, **55**: 62-73.
- Loessner, M.J. 2005. Bacteriophage endolysins-current state of research and applications. *Current opinion in microbiology*, **8**(4):480-487.
- Mead, P.S., Slutsker, L., Dietz, V., McCaig, L.F., Bresee, J.S., Shapiro, C., Griffin, P.M. and Tauxe, R.V. 1999. Food-related illness and death in the United States. *Emerging infectious diseases*, **5**(5): 607.
- Nath, S., Chowdhury, S., Sarkar, S. and Dora, K.C. 2013. Lactic acid bacteria–a potential biopreservative in sea food industry. *International Journal of Advanced Research*, **1**(6): 471-475.
- Ness, I.F. and Holo, H. 2000. Class II antimicrobial peptides from lactic acid bacteria. *Peptide Science*, **55**(1): 50-61.
- Oliveira, P.M., Zannini, E. and Arendt, E.K. 2014. Cereal fungal infection, mycotoxins, and lactic acid bacteria mediated bioprotection: from crop farming to cereal products. *Food microbiology*, **37**: 78-95.
- Piard, J.C., Muriana, P.M., Desmazeaud, M.J. and Klaenhammer, T.R. 1992. Purification and partial characterization of lacticin 481, a lanthionine-containing bacteriocin produced by Lactococcus lactis subsp. lactis CNRZ 481. *Applied and Environmental Microbiology*, **58(**1): 279-284.
- Pitt, J.I. and Hocking, A.D. 2009. The ecology of fungal food spoilage. In *Fungi and food spoilage* (pp. 3-9). Springer, Boston, MA.
- Rojas-Graü, M.A., Oms-Oliu, G., Soliva-Fortuny, R. and Martín-Belloso, O. 2009. The use of packaging techniques to maintain freshness in fresh-cut fruits and vegetables: a review. *International Journal of Food Science & Technology*, **44**(5): 875-889.
- Scholz, C.F. and Kilian, M. 2016. The natural history of cutaneous propionibacteria, and reclassification of selected species within the genus Propionibacterium to the proposed novel genera Acidipropionibacterium gen. nov., Cutibacterium gen. nov. and Pseudopropionibacterium gen. nov. *International journal of systematic and evolutionary microbiology*, **66(**11): 4422-4432.
- Silva, M.M. and Lidon, F.2016. Food preservatives–An overview on applications and side effects. *Emirates Journal of Food and Agriculture*, **1**:366-373.
- Stackebrandt, E.R.K.O., Cummins, C.S. and Johnson, J.L. 2006. Family propionibacteriaceae: the genus Propionibacterium. *The prokaryotes*, **3:** 400-418.

5

CHAPTER

ENZYMES FOR FOOD PRESERVATION

Insha Rasool*[1], Vikas Sharma[1], Nadira Anjum[2], Fozia Hameed[2], Anjum Ayoub[2] and Rukhsana Rahman[2]

[1]*Division of Biochemistry, Sher-e-Kashmir University of Agricultural Sciences and Technology- Jammu, Chatha-180009, India*

[2]*Division of Food Science and Technology, Sher-e-Kashmir University of Agricultural Sciences and Technology- Jammu, Chatha-180009, India*

Corresponding Author: Email ID: insha.ina20@gmail.com

ABSTRACT

Food processing involves the practices and approaches that are used to transform the raw materials of both plant as well as animal origin into products that are fit for human consumption. Enzymes on the other hand are basically protein molecules that act as catalysts thus increasing the rate of any chemical reaction. Enzymes play an important role in the field of food technology because of their ability to convert the raw materials into products. Moreover, enzymes act as preservatives that modify the properties of food. Enzymes are used in many food industries like meat, dairy, wine industry etc in the manufacture of pre-digested foods. The trend in food technology to avoid the synthetic food additives has forced the food technologists to use natural antioxidants and additives. Developing additional uses of enzymes in food processing requires more specificity as well as controlled conditions. Enzymes result in the production of compounds of interest either by hydrolysis of polymeric compounds or by their synthesis.

Keywords: Enzymes, Food processing, Food, Products

1. INTRODUCTION

Enzymes are defined as biological catalysts which increase the rate of reaction. They are actually proteins synthesized inside the bodies of all living organisms. Enzymes are highly specific in their mechanism of action *i.e.,* a specific enzyme acts on a specific substrate. Enzymes take part in all the life processes and metabolic activities (Hunter, 1995). There are two general classes of enzymes *viz.*, simple and conjugated.

(i) **Simple enzymes.** They are entirely proteins and consist of amino acids.

(ii) **Conjugated enzymes.** These enzymes consist of a protein part and a non-protein part. The non-protein part is called as cofactor/coenzyme.

A cofactor enhances the activity of an enzyme. It can be a simple metal ion or complex organic compound known as coenzyme. If a cofactor is removed from a conjugated enzyme it becomes inactive and is called apoenzyme. A complete and biologically active enzyme is called as holoenzyme.

$$\text{Apoenzyme} + \text{cofactor} = \text{Holoenzyme}$$

There are specific binding sites present on the enzyme molecules called as active sites at which the substrate binds (Robinson, 2015). There is a proper naming and classification of enzymes. Some enzymes have been named according to the type of substrate they are acting upon by adding the suffix -ase to the name of substrate e.g., urease hydrolyses urea, lipase hydrolyses lipids. While some enzymes acquired their names based on the type of reactions catalysed by these enzymes e.g., oxidases, reductases, transferases etc. However above two criteria do not hold true for some enzymes which prove as exceptions such as pepsin, trypsin etc. Hence, an International Commission on Enzymes was established called as Enzyme Commissionwhich classified enzymes according to six classes of reactions. Each enzyme is prescribed a four digit number, each of which denotes its major class, subclass and sub-subclass (Robinson, 2015). The last digit is a serial number in the sub-subclass. This is known as EC number. The six classes of enzymes are;

(i) EC 1 Oxidoreductases

(ii) EC 2 Transferases

(iii) EC 3 Hydrolases

(iv) EC 4 Lyases

(v) EC 5 Isomerases

(vi) EC 6 Ligases.

Enzymes work through a course of three main phases. The enzyme first combines with substrate at the active site forming the enzyme-substrate complex and then the product is finally formed.

$$E + S \rightarrow \text{ES complex } E + P$$

There are several factors which determine the enzyme activity like temperature, pH, substrate concentration,enzyme concentration etc. But enzyme activity is affected mainly by temperature and pH (Vitolo, 2020).

Food processing is a set of practices that are followed in the transformation of raw food materials into food products that we eat.(Monteiro and Levy,2010).

There are different degrees or extent up to which food materials can be processed ranging from minimally processed foods to highly processed foods (Ohlsson and Bengtsson,2002); Monteiro et.al ,2011).Food processing adversely affects the quality of foods and for every particular food there is a particular process. There is a number of processing methods that can add value to our foods and can improve their nutritional benefits. Some of the primitive methods implemented for food processing include cooking, drying, fermenting and preserving by adding salt or other preservatives. While some of the recent food processing methods include homogenization, pasteurization, termination, sterilization, blanching, deep freezing, use of enzymes, etc. As far as food preservation is concerned, it is an ancient practice to preserve food from spoilage by microorganisms with the help of food preservatives. Food preservatives are chemicals that increase the shelf life of food products and preserve their natural characteristics. Also food preservatives play an important role while the foods are being transported overseas or being stored from season to season (Seetaramaiah et al., 2011).In addition to the use of preservatives there are some modern packaging techniques as well e.g., vacuum and hypobaric packaging. However this all aims at preserving not only the taste and aroma of food but also preserve the natural appearance of food for a longer period of time. There is evidence of using various phytochemicals as natural antimicrobial agents which would be safe for human consumption e.g., *Azadirachtin* from Neem, menthone from mint (Ward et al., 1998).There are three mains classes of food preservatives and they are;

(i) Natural preservatives: are of natural origin and cause no harm to our health. These include salts, sugars, oil, vinegar etc.

(ii) Artificial preservatives: are chemically synthesized substances that are used to increase the shelf life of food products and decreases chances of microbial contamination e.g., benzoates, nitrites etc.

(iii) Microbial preservatives : are preservatives that inhibit the growth of microorganisms like bacteria or fungi and avoid food oxidation. e.g., Methyl, ethyl parabens, Sorbic acid, BHA, ethanol etc (Seetaramaiah, *et al.*, 2011).

2. APPLICATION OF ENZYMES IN FOOD PROCESSING AND FOOD PRESERVATION

Various enzymes have been used in the food processing and food preservation since ancient times e.g., in wine and beer making, bread and cheese making etc. Later on large number of enzymes were isolated from various microorganisms and living cells. Then they were produced and used commercially on a large-scale in the food industry. Although microorganisms are believed to be the best sources of commercial enzymes in today's world as they can produce huge amounts of the desired enzymes (Agarwal and Sahu, 2014). Since enzymes can convert completely raw materials into improved food products and can modify their characteristics and also increase nutritional value of ingredients and food products, enzymes have a great importance in food industry (Ward and Moo-Young, 1988). Due to their stability and consistent production microbial enzymes are much preferred over plant and animal enzymes. Also they prove to be less costly and easily optimized. Enzymes of bacteria, fungi and yeast are used in food preparations to improvise taste and quality which has led the food industries to huge economic benefits. Several microorganisms have been genetically modified with the use of biotechnology and are grown commercially in technical fermenters in order to incorporate desired characteristics and modifications. Most of the enzymes used in food technology are hydrolytic in nature breaking/converting complex molecules into simpler ones e.g., lipases, amylases, proteases, carbohydrases etc.

2.1. ENZYMES IN BAKING INDUSTRY

Baking is an important part of day to day life as we all know how popular the bakery is in today's world scenario. Bread and bakery have become a daily need of all the households because of the taste and easily digested. In fact, bread has become the staple food in many countries and no doubt bakery is incomplete without bread. Enzymes are important ingredient used in most of the baking procedures in order to preserve the freshness and enhance the quality of bakery products. Since modern world prefers organic over synthetic chemicals, enzymes have achieved a great importance in baking industry especially in bread making (Cauvain and Young, 2006).There are three main sources of enzymes in baking; endogenous enzymes present within the flour, enzymes associated with microbes present in the flour and exogenous enzymes added deliberately to the dough (Di Cagno *et al.*, 2003). Usually enzymes are added in the dough in order to modify the rheology of dough, to maintain gas retention and softness of the crumb in bread making and enzymes also play a role in maintaining the texture and softness of dough in pastry, cake and biscuit manufacturing and reducing the formation of acrylamide in bakery products (Cauvain and Young, 2006).Enzymes act as technological aids in baking processes and are usually added to the flour at the mixing step of bread making. The most commonly used enzymes in bread making process are -amylases (Sanz Penella *et al.*, 2008). Amylases breakdown starch into smaller and simpler products so that they are easily acted upon by the yeast. Other enzymes such as hemicellulases, xylanases, oxidases and lipases help in improving the quality of products.

Table 1: Role of various enzymes in bakery industry

S.no	Enzyme	Role
1.	α-amylase	Breakdown of starch in flour, maintaining volume and softness of bread crumb
2.	Oxidoreductase	Increase the strength of gluten
3.	Proteases	Lowering of protein content in flour

4.	Lipases	Improvisingthe gas retention in dough
5.	Lipoxygenase	Dough bleaching and strengthening
6.	β-xylanases	Maintaining dough stability and improving dough handling
7.	Glucose oxidase	Carries out oxidative reaction with gluten to make weaker doughs more elastic and strong
8.	Asparaginase	Preventing the formation of acrylamide during baking
9.	Maltogenic α-amylases	Improving the shelf-life of bread and other baked products

Source: Bloom *et al.*, **2005; Riberiro** *et al.*, **2010**

2.2 ENZYMES IN DAIRY INDUSTRY

Dairy industry in India is a well established sector, India being the largest producer of milk in the world. Milk is regarded as a complete food because of its nutritional value and is an essential in daily life. While there is a surplus availability of milk in India it urges the dairy industry to meet the demands of people by manufacturingvalue added products from milk like curd, yoghurt, cheese etc. This can only be done by carrying out various biochemical processes involving enzymes. Use of exogenous enzymes like rennet in cheese making is the earliest technique used in food processing or dairy industry. Proteinases have also been used nowadays in dairy industry for increasing the rate of cheese ripening process, modifying the characteristics and preparation dietic products. In addition to this animal rennet has played a great role in improving the quality and texture of cheese by acting as an excellent milk-coagulant (Bhoopathy, 1994). Milk contains lactose sugar and for breaking this sugar a hydrolase enzyme is needed and that enzyme is lactase or β-galactosidase obtained from plants, animal organisms, yeasts and bacteria (Mehaiya, 1987). Among all the enzymes used in dairy industry, proteases and lipases have a significant importance while minor enzymes having little use in dairy industry include glucose oxidase, catalase, superoxidedismutase, sulphydryl oxidase, lactoperoxidase, and lysozymes. Enzymes used together in dairy food preservation are usually glucose oxidase and catalase.

Table.2: Role of enzymes used in dairy industry

S.no.	Enzyme	Role
1.	β-galactosidase (lactase)	Hydrolyse lactose into glucose and galactose, sweetening of whey and milk products, improve functionality of whey protein, avoid crystallisation in ice cream of whey etc.
2.	Lactoperoxidase	Helps in cold sterilization of milk
3.	Chymosin	Hydrolyses casein and helps in coagulation of milk for cheese making

4.	Neutral proteinases	Increase the rate of cheese ripening, improve quality of cheese and cheese products
5.	Lipase	Hydrolyse triglycerides to fatty acids and glycerol, improve the quality of cheese products by flavour enhancement
6.	Aminopeptidase	Releases free amino acids, debittering and catalysing the rate of cheese maturation

Source: Whitehust and Law, 2002)

2.3 ENZYMES IN MEAT PROCESSING

Meat in our diet holds a special place and is considered as an important component of a meal. It is a luxurious commodity because it is way too expensive than all the other foods so, it becomes necessary to check the quality being provided and select proper cuts at the butcher's shop. The most important factor upon which quality is mainly evaluated is the tenderness of meat (Boleman*et al.*,1995; Koohmaraie and Geesink, 2006) and it is determined by the management of pre and post slaughtering handling practices carried out. The tenderness of meat is due to the endogenous enzyme activity and loosening of connective tissue while the toughness of meat being attributed mainly to the presence of collagen, actomycin and elastin (Destefanis*et al.*, 2008). This tenderness can be increased by the application of several processing techniques known as tenderizing or tenderization of meat. Toughness of meat cuts can be overcome by tenderizing with the use of enzymes like proteases. For example enzymes like papain and bromelain have been extensively used for tenderization of meat, derived from plant sources papaya and pine apple in order to tenderize the meat and retain its freshness (Destefanis*et al.*, 2008).

Table 3: Enzymes used in meat processing

S.no.	Enzyme	Role
1.	Glutaminase	Enhance flavour of the meat due to L-glutamic acid.
2.	Tyrosinase	Cross-linking of meat proteins, enhancement of functional properties of enzymes.
3.	Acid proteases	Improve flavour, nutritional andfunctional characters of proteins, convert animal carcasses into flavorous compounds
4.	Papain/ficin/bromelain	Maintaining tenderness of meat, Increase palpability, solubility and digestibilityof meat protein.
5.	Transglutaminase	Improve the quality and structural properties of processed or cooked meat.
6.	Lipase	Improve the flavour in sausages
7.	Elastase	Promotes tenderization of meat, improve the market value of low value meat.
8.	Actidin	Improve tenderness of cooked or processed meat.

Source: Bloom et al., 2005

2.4 ENZYMES IN BREWING INDUSTRY

Brewing industry is one of the world's leading industries and beer is world's most consumed beverage (Nelson, 2005). Beer is prepared from barley and requires several enzymes for its fermentation by yeast. Before the action of yeast, the sugar of barley that is starch needs to be broken first into simpler fermentable sugars so that it gets easier for the yeast to act on them and form alcohol. This is done bythe enzyme -amylase which carries out the saccharification of starch. Basically there are four main processes according to which the enzymes are used for the brewing process such as germination, mashing, fermentation and clarification (Bamforth, 2009). Beta glucanase, protease, alpha amylase, and beta amylaseare the most common enzymes used in brewing industry(Aastrup*et al.*, 2004).

- α-amylases- for saccharification of starch, can be bacterial and fungal -amylases.
- β-glucanase- for breakdown of barley β-1,4- and β-1,3- linked glucan.
- Neutral protease- for the hydrolysis of protein to increase the fermentation rate later.

Other enzymes such as cellulases, papain, xylanases etc are also used in brewing industry.

Table 4: Role of enzymes in brewing industry

S.no.	Enzyme	Role
1.	α-amylases	Hydrolyse starch to reduced viscosity,increase maltoseand glucose content
2.	β-glucanases	Hydrolyse glucans into oligomers leading to lower viscosity and better filterability; Improve wort separation.
3.	Proteases	Improve yeast growth. Improvement of malt, increase soluble protein and free amino-nitrogen (FAN).
4.	Amyloglucosidases	Increase glucose content, Increase 1% fermentable sugar in light beer.
5.	Pentosanases, xy-lanases	Hydrolyze pentose sugars of barley wheat and malt, improve beer extraction and filteration
6.	Pullulanases	Hydrolyse α-1, 6 branching linkages of starch or saccharification of starch Ensure complete fermentability of the Wort.
7.	α-acetolactatedecar-boxylases (ALDC)	Maintaining the taste and flavour, Convert α -acetolactate to acetoin directly, Decreasing fermentation time by preventing formation of diacetyl.

Source: Whitehust and Law, 2002

2.5 ENZYMES IN FRUIT AND VEGETABLEPROCESSING AND JUICE EXTRACTION

Fruits and vegetables are a basic necessity in day-to-day life. Fruit juices are considered as important part of a healthy diet plan. We depend on fruits and vegetables as they are rich sources of vitamins and other important nutrients so it becomes necessary to preserve the quality and ensure high quality of juices produced from these fruits and vegetables (Bhardwaj and Pandey, 2011). Here enzymes play a very vital roleas in processing of fruits and vegetables as well as increasing their post-harvest shelf life. India is the second largest producer of fruits and vegetables in the world market. It becomes important to aid the processing techniques which includes use of enzymes for a number of benefits(Kumar, 2015). Enzymes can thoroughly enhance the quality of juices and their clarity, ensure improved concentration, increase the quantity of yield, less wastage, improve the essential oil extraction from peel etc (Grassin and Fauquembergue, 1996). The examples of enzymes mostly used in fruit and vegetable processing industry are mainly pectinases and then amylases, cellulases, hemicellulases, glucoamylases etc. (Lee *et al.*, 2006; Sandri *et al.*, 2012).

Table 5: Role of enzymes in fruit and vegetable processing

S.no.	Enzyme	Role
1.	Amylase, glucoamylase	Breakdown of starch into glucose, clarification of cloudiness in juices
2.	Pectinase	Breakdown pectins present in cell walls, improve juice production
3.	Cellulase, hemicellulase	Accelerate pectin hydrolysis, Lowering viscosity, maintenance of texture, fruit liquifaction in juice Production
4.	Laccase	Increase the susceptibility of browning in storage
5.	Naringinase and limoninase	Avoid bitterness by acting on compounds that cause bitterness in citrus fruits

Source: Fernandes, 2010

2.6 ENZYMES AS FOOD ANTIOXIDANTS

Food antioxidants are compounds that scavenge and neutralise the free radicals generated by the random oxidation of foods. They are highly important food additives for they help to prevent the spoilage of fatty foods from oxidative rancidity and preserve the flavour, colour and aroma of stored foods.(Kebede and Admassu, 2019). Fat based foods such as dairy products, meat and fried foods are at a risk of lipid peroxidation so there is a need to use food antioxidants which can be natural or synthetic. Enzymes act as natural food antioxidative agents and prevent the oxidation of food by promoting three types of reactions;

1. Removal of oxygen
2. Removal of free radicals
3. Reduction of lipid hydroperoxides

Examples of enzymes as food antioxidants are glutathione oxidase, glucose oxidase, catalase and superoxide dismutase. Glucose oxidase which is an oxidoreductase enzyme extracted from the fungus *Aspergillus niger* used widely in food industries(Chaudhary et al, 2015). This enzyme catalyzes the conversion of glucose into gluconic acid using oxygen as an electron acceptor resulting in the formation of hydrogen peroxide. This hydrogen peroxide is then acted upon by catalase enzyme. The studies show that lowering the concentration of oxygen and peroxides inhibits the generation of free radicals and thus decreasing the rate of lipid peroxidation (Meyer and Isaken, 1995).

2.7 IMMOBILIZED ENZYME TECHNOLOGY FOR FOOD APPLICATION

An immobilized enzyme is an enzyme which is attached to neutral or inert insoluble organic or inorganic carrier molecules like calcium alignate or silica so as to increase their resistance to the external environmental changes such as temperature or pH (Homaei *et al.* 2013). Enzyme immobilization ensures many benefits such as multiple or repetitive use of same batch of enzymes, increase in stability, easy separation of enzyme from the productetc. There are several techniques used for the immobilization of enzymes which include adsorption, entrapment and encapsulation, covalent or cross linking (Sheldon, 2007).

Table 6: List of immobilised enzymes used in food industry

Enzyme	Immobilization support	Food substrate	References
β-galactosidase and amyloglucosidase	Bone powder, Duolite A-568, other organic and inorganic molecules	Lactose, Muscat wine, whey, skimmed milk	Husain *et al.*, 2010, Carpio *et al.*, 2000
Glucoamylase	Chitin	Starch and itshydrolysed-derivatives	Freire andsant'An-na,1990
Trypsin	Cellulose	Milk protein like β-lactoglobulin	Yamamoto *et al.*, 2005
Pectinase	Anion exchange resin	Pectin	Sariogluet *al.*, 2001
Lipase	Calcium-alginate beads	Fat and oil	Jeganathanet *al.*, 2006
Tyrosinase	Polyacrylic-acid carbon nanotubes	Phenolic compounds present in wine	Kim *et al.*, 2010

FUTURE PERSPECTIVES AND CONCLUSION

The use of enzymes in food processing and preservation has proved to be a reliable and highly beneficial approach because of their potent catalytic action and no side effects. They are completely natural agents and thus save the energy, environment and resources. Since enzymes have diversified roles in every type of food processing because of their specificity and quick effectivity, there is a scope for increasing and improving this utilization and diverse application of enzymes everywhere. Enzymes act as technological tools hence, constant efforts are being made to design these enzyme systems according to the convenience of application. There is a scope for improvement of basic properties such as more stability, less susceptibility to the action of enzyme inhibitors, less dependence on cofactors or metal ions and more effective mechanism of action. Immobilization of enzymes has proved to be an effective tool in making enzymes suitable for application in food industry. Inspite of the advantages it has led to, like multiple or repetitive use of a single batch of enzymes, more stability, easy separation of enzyme and product etc, there is still something lacking in this field e.g., proper use and selection of method for enzyme immobilization, selection of a proper carrier etc. In today's world of increasing population, increasing demands and exhausting natural resources, enzyme technology can greatly help the food processing industries to meet the challenges and fulfill the needs of this growing population.

REFERENCES

- Agarwal, S. and Sahu, S. 2014. Safety and Regulatory Aspects of Food Enzymes: An Industrial Perspective. *International Journal Interdisciplinary and multidisciplinary Studies,* **1**: 253-267.
- Bamforth, C.W. 2009. Current perspectives on the role of enzymes in brewing. *Journal of Cereal Science,* **50**: 353-357.
- Bhardwaj, R.L. and Pandey, S. 2011.Juice blends- a way of utilization of inder utilized fruits, vegetables, and spices: a review *Critical Reviews in Food Science and Nutrition,* **51**(6): 563-570.
- Bhoopathy, R. 1994. Enzyme technology in food and health industries. *Indian Food Industry,* 13: 22–31
- Bloom, J. D., Meyer, M. M., Meinhold, P., Otey, C. R., MacMillan, D. and Arnold, F. H. 2005. Evolving strategies for enzyme engineering. *Current Opinion Structural Biology,* **15**: 447-452.
- Boleman, S. J., Boleman, S. L., Binder,T. D., Mc Millin, K. W. And Monlezun, C. J. 1995. Effect of post-mortem time of calcium chlorideinjection on beef tenderness and drip,sooking and total loss. *Meat Science,* **39**: 35-41.

- Caprio, C., Gonzalez, P., Ruales, J. and Bastista, F. 2000. Bone-bound enzymes for food industry application. *Food Chemistry*, **68**: 403-409.
- Cauvain S. and Young L.2006.Ingredients and their influences. In: Cauvain S, Young L. (eds.) Baked Products. Science, Technology and Practice. Oxford: Blackwell Publishing.72-98.
- Chaudhary, S., Sagar, S., Kumar, M., Sengar, R.S. and Tomar, A. 2015. The use of enzymesin food processing: A review. *South Asian Journal of Food Technology and Environment*, **1**(3&4): 190-210.
- Destefanis, G., Brugiapaglia, A., Barge, M. T. and Dal Molin, E. 2008. Relationship betweenbeef consumer tenderness perception and Warner-Bratzler shear force. *MeatScience*, **78**: 153-156.
- Di Cagno, R., De Angelis, M., Corsettic, A., Lavermicocca, P., Arnault, P., Tossut, P., Gallo,G. and Gobbetti, M. 2003. Interactions between sourdough lactic acid bacteria and exogenous enzymes: effects on the microbial kinetics of acidification and dough textural properties. *Food Microbiology*, **1:** 67–75.
- Fernandes, P. 2010a. Enzymes in Food Processing: A Condensed Overview on Strategies for BetterBiocatalysts. *Enzyme Research*, 2010:862537.
- Freire, D. G. and Sant'Anna Jr., G. L.1990. Characterization of a glucoamylase immobilized on chitin. *Biomass*, **23:** 71-78.
- Grassin, C. and Fauquembergue, P. 1996. Fruit juices. In: Godfrey, T., West, S. (Eds), Industrial Enzymology (2nd edn). *Macmillan*, London. pp. 227.
- Homaei, A., Sariri, R., Vianell, F. And Stevanato, R. 2013. Enzyme immobilization: an update. *Journal of Chemical Biology* **6**: 185–205.
- Hunter, T. 1995. Protein kinases and phosphatases: the yin and yang of protein phosphorylation and signalling. *Cell*, **80**: 225–236.
- Husain, Q. 2010. Beta galactosidase and their potential application: A review. *Critical Reviews in Biotechnology*, **30**: 41-62
- Jeganathan, J., Bassi, A. and Nakhla, G. 2006. Pre-treatment of high oil and grease pet food industrial wastewaters using immobilized lipase hydrolyzation. *Journal of HazardousMaterials*, **137**: 121-128.
- Kebede, M. And Admassu, S. 2019. Application of antioxidants in food processing industry: Options to improve the extraction yields and market value of natural products, *.Advances in Food Technology and Nutrition Sciiences*, **5**: 38-49.
- Kim, O., Lee, J. C., Robards, K. and Choi, S. H. 2010. Immobilization of tyrosine in carboxylic and carbonyl group-modified MWNT electrode and its application for sensing phenolics in red wine. *Journal of Nanoscience and Nanotechnology*, **10**: 3790-3798.

- Koohmaraie, M. and Geesink, G. H. 2006. Contribution of post-mortem muscle biochemistryto the delivery of consistent meat quality with particular focus on the calpain system. *Meat Science,* **74**: 34-43.
- Kumar, S. 2015. Role of enzymes in fruit juice processing and its quality enhancement. *Advances in Applied Science Research,* **6** :114-124
- Lee, W. C., Yusof, S., Hamid, N. S. A. and Baharin, B. S. 2006. Optimizing conditions for enzymatic clarification of banana juice using response surfacemethodology (RSM). *Journal of Food Engineering,* **73**: 55-63
- Mehaia, M. A. and Cheryan, 1987. Production of lactic acid from sweet whey permeate concentrates. *Process Biochemistry,* **22**: 185–188.
- Meyer, S. and Isaksen, A. 1995. Application of enzymes as food antioxidants, *Trends in Food Science &Technolog,***6:** 300-304.
- Monteiro, C.A., Levy, R.B., Claro ,R.M., Castro de and Cannon, G.2011. Increasing consumption of ultra-processed foods and likely impact on human health. *Public health nutrition,* **14:**5-13,
- Monteiro,C.A. and Levy, R.B. 2010. A new classification of foods based on the extent and purpose of their processing , *Public Health Nutrition,* **26:**2039-2049.
- Ohlsson ,T. And Bengtsson ,N. 2004. Minimal Processing Technologies in the Food Industry. Boca Raton, FL: CRC Press.
- Riberiro D. S., Henrique, S. M. B., Oliveira, L. S., Macedo, G. A. and Fleuri, L. F. 2010. Enzymes in Juice processing: A review. *International Journal of Food Science andTechnology,* **45**: 224-230.
- Robinson, P.K. 2015. Enzymes: principles andbiotechnological applications. *Essays in Biochemistry,* **59**:1-41.
- Sandri, I. G., Fontana, R. C., Barfknecht and Da Silveira, M. M. 2012. Clarification of fruit juices byfungal pectinases. *Food Science and Technology,* **44**: 2217-222
- Sanz Penella, J. M., Collar, C. and Haros, M. 2008. Effect of wheat bran and enzyme addition on dough functional performance and phytic acid levels in bread. *Journal of CerealScience,* **48:** 715–721.
- Sarioglu, K., Demir, N., Acar, J. and Mutlu, M. 2001. The use of commercial pectinase in the fruit juice industry. Part 2: Determination of the kinetic behaviour of immobilized commercial pectinase. *Journal of Food Engineering,* **47**: 271-274.
- Seetaramaiah, K., Smith,A. A., Murali, R., Manavalan, R. 2011.Preservatives in Food Products – Review. *International Journal of Pharmaceutical & Biological Archives,* **2**: 583-599.

- Sheldon, R. A. 2007. Enzyme immobilization: the quest for optimum performance. *Advanced Synthesis and Catalysis,* **349**: 1289– 1307.
- Vitolo, M. 2020. Brief Review on Enzyme Activity. *World Journal of Pharmaceutical Research,* **9**: 60-76.
- Ward, O. and Moo-Young M. 1988. Thermostable enzymes. *Biotechnology Advances,* **6**: 39-69
- Ward, S.M., Delaquis, P.J., Holley R.A. and Mazza G. 1998. Inhibition ofspoilage and pathogenic bacteria on agar and pre-cooked roasted beef byvolatile horse radish distillates. *Food Research International,* **31:**19 26.
- Whitehust, R. J. and Law, B. A. 2002. The nature of enzymes and their action in foods, In: Enzymes in Food Technology. *Sheffield AcademicPress,* UK.
- Yamamoto, S., Imamura, A., Susanti, I., Hori, K., Tanji, Y. and Unno, H. 2005. Effect of spacer length on beta-lactoglobulin hydrolysis by trypsin covalently immobilized on a cellulosic support. *Food and Bioproducts Processing,* **83**: 61-67.

CHAPTER 6

BASIC AND APPLIED CONCEPTS OF EDIBLE PACKAGING FOR FOOD PRESERVATION

[1]Chavan R. F.*, [2]Sakhale B. K. and [3]Anjum Ayoub

[1,2]Department of Chemical Technology, Dr. Babasaheb Ambedkar Marathwada University, Aurangabad (MH)-431004, India

[3]Division of Food Science and Technology,Sher-e-Kashmir University of Agricultural Sciences and Technology, Jammu (J&K)-180009, India

**Corresponding author: ramesh.chavaan@gmail.com*

ABSTRACT

The edible coatings or films have received extensive consideration lately due to lack of investigation regarding its safety issues for using in food applications. The binding materials utilized in edible packaging such as polysaccharide, lipid and protein-based materials and their composites have evident focal points over engineered edible film coatings. This would help in decreasing of ecological contamination and damage that may have incurred by traditional plastic based packaging materials. The primary and important advantage of using these consumable films over other engineered materials is that these might be securely eaten as a piece of the food items along with the foods packed contained with and are climate inviting simultaneously as they increment the time frame of realistic usability of new food items. By working as boundaries, palatable films possibly lessen the unpredictability and subsequently improve the recyclability of bundling materials, contrasted with the more conventional non-ecological agreeable bundling materials, and have the possibility to substitute manufactured polymer films. Plasticisers are for the most part added to palatable films for upgrade in adaptability and solidness. Cell reinforcements are added to edible bundling materials to postpone the pace of oxidation responses. One of the huge arising elements of edible coatings is their utilization as transporters of antifungal and antimicrobials specialists to improve the timeframe of realistic usability of nourishments items; they likewise utilized as transporters of supplements to expand the healthy benefit of last prepared food items. Likely uses of consumable edible films as keen packaging needs research, which is as yet in its earliest stages and augmentation on its mechanical use, will be an achievement in the packaging business.

Keywords: Edible packaging, preservation, shelf life, coating, films, nutrients

1. INTRODUCTION

33% of the delivered food prompts squander yearly particularly during transportation and capacity because of physicochemical, microbial waste, oxidation, dampness changes and different components (FAO, 2011; Hannon et al., 2017). The Rotting of food causes misfortune as well as weakening in wholesome and sensorial quality, which prompts diseases or may even prompt harmfulness (Gómez et al., 2014; Hadian et al., 2017). Preferred position of Bundling of food handling after the cycle is finished, empowering nourishments to ship securely for significant distances from their cause point and still be nutritious at the hour of utilization (Ahmad et al., 2017; MacGowan and Macnaughton, 2017; Nordstrom and Malmsten, 2017; Ribeiro et al., 2017; Lin et al., 2018). A urgent capacity of food bundling is recognizability, comfort, and alter sign. It assists with making food is savvy so that fulfills industry necessities and purchaser wants, keeps up food quality, security, and diminishes ecological effect (Swamp and Bugusu, 2007).

In any case, food bundling innovation keeps up food quality with different issues, including material expenses and energy; improve natural and social awareness and exacting guidelines on contaminations and removal of metropolitan strong waste. Consumable movies and coatings are one of the arising approaches for food-quality streamlining. Consumable coatings and movies have gotten extensive consideration as a result of their favorable circumstances contrasted with engineered films. One of the critical advantages of utilizing palatable bundling frameworks, when contrasted with manufactured bundling frameworks, is that they are a vital piece of the food item; they can be eaten, without unload and toss the bundle (Bourtoom, 2008; Ali et al., 2010; Trinetta, 2016). The expanded interest in eatable coatings has been inspired by expanding purchaser interest for protected, helpful, and stable nourishments, and the familiarity with the unfriendly ecological impacts of non-biodegradable bundling (Bourtoom, 2008; Donhowe and Fennema, 1993.

The helpfulness of it depends on the ability to keep up the quality, to build the time span of usability, and to add to the monetary proficiency (Arismendi et al., 2013). New kind of bundling materials have been created and portrayed by researchers from characteristic assets; yet in spite of that, this data is accessible for the arrangement of food wrapping, it isn't all inclusive for all items, which represents a test for the advancement of explicit coatings and movies for every food (Aguirre et al., 2018). The bundle ought to fulfill different prerequisites successfully and monetarily; thus, another food bundle should be improved and coordinated viably with the food inventory network. Changes in the technique for appropriation chain work, for example, the creation, dispersion, put away and retailed items, mirror the proceeding with increment for these materials. As of late, Palatable wrapping give a substitution as well as stronghold of the characteristic layers at the external surfaces of item to forestall dampness misfortunes, gas smells and solute developments out of the food, while specifically considering controlled trade of basic gases, for example, carbon dioxide, oxygen, and ethylene, which are engaged with food item breath (Embuscado and Huber, 2009). Besides, the materials that are utilized for this reason can altogether cover the food or can be utilized as a ceaseless layer between food segments (Guilbert, 1995).

2. NATURAL POLYMERS BASED EDIBLE FILMS AND COATINGS

Current purchaser requests and needs about more characteristic, high-caliber, and more secure nourishments around the globe have changed the worldwide market. All things considered, they likewise request food bundles that don't build contamination, and are made by reasonable cycles, these in a modest way. Hence organizations and scientists have zeroed in on creating supportable, biodegradable, and eatable materials that improve the efficiency, food quality, renew, and give sanitation (Mahalik and Nambiar, 2010). Biopolymers have been read for analysts as an option in contrast to conventional (oil) food bundling with respect to their film-development properties to deliver biodegradable and eatable movies and coatings for food bundling (Azeredo et al., 2009). Biopolymers figure eatable movies and coatings are organizing ones, for example, polysaccharides, proteins, and lipids (Espitia et al., 2014a).

A definition for eatable films and coatings is that they are essential bundling produced using consumable materials. Additionally, it is conceivable to apply a dainty layer of consumable bundling straightforwardly in the food by inundation, showering, and soaking (covering) or by being recently shaped into a film and after are utilized as a food wrap without changing the cycle strategy for the covering and the fixings utilized (Galus and Kadzi ska, 2015). So the distinction between a consumable film and covering is that coatings are applied in fluid structures while films are acquired as strong covers and afterward applied to food stuff (Falguera et al., 2011).

2.1 POLYSACCHARIDE-BASED EDIBLE FILMS

Polysaccharides, like cellulose, alginate, carrageenin, gum xantan, and starch, are employed in recent years as biopolymer compounds to make edible films and coatings so as to scale back ancient plastic packages and are applied on food surfaces as shown in Figure 1 (Espitia et al., 2014b). one in all the materials that are recently used as a property compound for edible film formation area unit lignocellulosic ones (Mellinas et al., 2016). Next we tend to describe a number of the foremost used polysaccharides in edible films and coatings formulation.

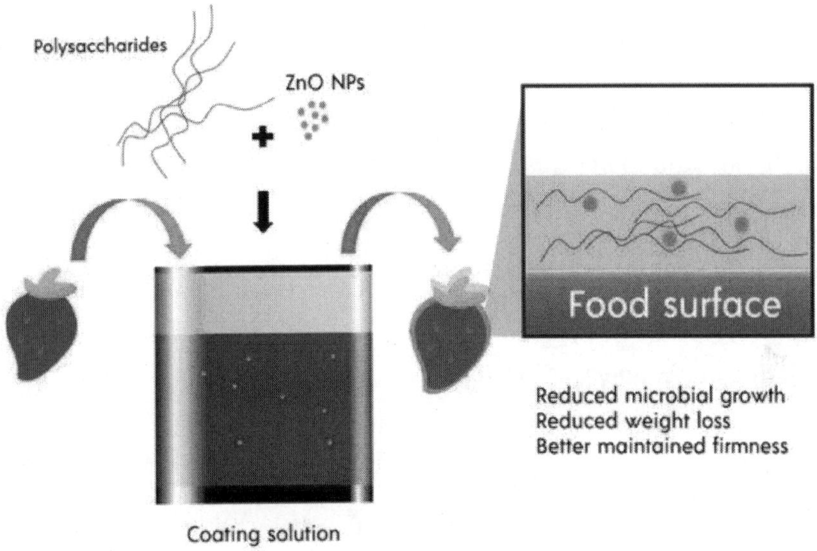

Figure1. Process of polysaccharide based coating on food surface

Characteristics of polysaccharides area unit that they're not cyanogenetic and knavish accessible compounds in nature, and have selective porosity to atomic number 8 and carbonic acid gas (Erginkaya et al., 2014). This characteristic permits polysaccharide-based edible films and coatings to prolong time period of fruits, however carbohydrate exclusively packages have the disadvantage of tide porosity.

2.1.1 Animal origin polysaccharides

Chitin and chitosan. Chitin is discontinuance cellulose, the in a holding pattern beat on all occasions going on biopolymer in monogram. It is debased in the exoskeleton of crustaceous, in fungal chamber walls, and backup coordinated information. By the deacetylation of chitin, chitosan is second-hand [poly-β-(1 4)-N-acetyl-dglucose- amine], a clever accessory of the shells of crustaceans. Chitosan is a contemptuous molecular level cationic polysaccharide back present-day antibacterial and antifungal activities, as fully as top-drawer film-forming abilities (Campos et al., 2011; Ferreira et al., 2009).

2.1.2 Plant origin polysaccharides

Cellulose is that the major part of the plant cell, therefore it's the foremost abundant chemical compound on earth. it's shaped by d-glucose units joined through β-1,4 organic compound bonds. polyose derivatives square measure primarily wont to kind natural, perishable, or edible films as they're tasteless, odorless, and perishable substances with a coffee application value (Erginkaya et al., 2014). Most used polyose derivatives square measure carboxymethylcellulose (E466, CMC), methylcellulose (E461, MC), and hydroxypropylmetylcellulose (E464, HPMC) (Çag˘rı et al., 2002). Starch could be a sugar composed of enzyme (25%) and amylopectin (75%) (Bourtoom, 2008) wide obtainable in nature and is made to cloth perishable films as starch films square measure clear or clear, flavorless, colorless, and tasteless (Skurtys et al., 2011). Pectin. Pectins square measure a bunch of plant-derived polysaccharides found in fruits and vegetables, the bulk extracted from citrus peel and apple pomace (Dhanapal et al., 2012). cellulose (E440) is associate anionic sugar with structural backbone of (1 4)-linked -dgalacturonic acid unit and employed in food as gelling, helpful, and thickening agent in product, like yogurts, jams, milk, and ice-cream (Espitia et al., 2014a,b).

2.1.3 Marine origin polysaccharides

Alginates ar a natural carbohydrate extracted from the marine alga from the family Phaeophyceae; they're combined by units of R-d-mannuronate (M) and a-l-guluronate (G) at totally different ratios and distributions within the chain (1–4); generally the sequences of M and G within the chain depends on the supply of alginate and therefore the age of the plant (Erginkaya et al., 2014). Formation of gels by the addition of metal ions involves the G blocks, that the higher concentration of G units the upper gel strength (Albert et al., 2010). however, alginate-based coatings will gift smart quality and preserve food period of time by increasing the water barrier, maintaining the flavour, and retarding fat oxidation; they'll be used as carriers for imicrobals and antioxidants to attain a high concentration of preservatives in foods (Song et al., 2011).

Another marine origin carbohydrate is that the gum that ar sulfated water soluble polymers extracted from numerous red seaweeds of the class family. they're utilized in food, dairy, and pharmaceutical industries, like gelling, emulsifying, and helpful ingredients (Seol et al., 2009). 3 major sorts of carrageenans are: k, , and -carragenans, looking on the amount and position of sulphate teams that respectable ar twenty, 33, and forty first (w/w) (Fabra et al., 2008). Karbowiak et al. (2006) reports that the mechanism of gum film formation includes gelation throughout moderate temperature drying, leading at solid film fashioned by polysaccharide-double helices once solvent evaporation.

2.1.4 Microbial polysaccharides

Gellan may be a category of polysaccaride made by the microorganism Sphingomonas genus Elodea (also referred to as Pseudomonas elodea) and presents distinctive mixture and gelling properties and smart ability to create coatings (Moreira et al.,

2015). the employment of gellan within the food trade is increasing wherever it's used as a texturizing and gellin agent (Rojas-Graü et al., 2008) and additionally as a carrier for food additives, like antibrowning and antimicrobial agents, colorants, flavors, and nutraceuticals (Oms-Oliu et al., 2010; Robles-Sanchez et al., 2013).

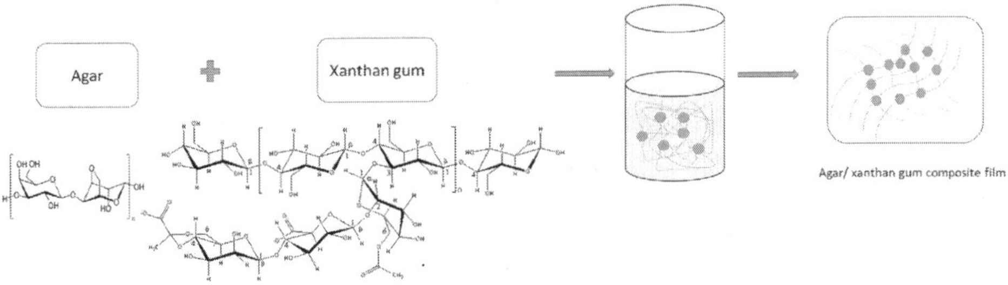

Figure 2. Xanthan gum and agar based composite film

Xantan gum is associate degree exopolysaccharide synthetized by the microorganism bacteria genus campestris, may be a usually recognized as safe (GRAS) compound (FDA, 2013) as food stabilizer, thickener, and surfactant and used for making edible films along with agar as shown in Figure 2. The viscous answer that it forms in cold or plight is stable at a high varies of hydrogen ion concentration and temperature and is also stable to accelerator degradation (Sharma and Rao, 2015). It's a structure of one,4-linked β-d-glucose residues associate degreed a facet chain of oligosaccharide guaranteed to an alternating d-glucose residues. The oligosaccharide chain is created is created acid—1-2- -d-mannose (Zambrano-Zaragoza et al., 2014). Xantan gum-based edible coatings are used recently to enhance quality and period, additionally as a carrier of bioactive compounds of minimally processed prickly pear (Mohamed et al., 2013) and contemporary cut apples (Freitas et al., 2013), among different fruits.

3. LIPID-BASED EDIBLE COATINGS

Lipids square measure compounds that have the aptitude for mixable or nonionic organic solvents, however few of them contain hydrophilic and hydrophobic half forming particle (Fig. 1.1). the variety of the cluster is formed up by monoglycerides, diglycerides, triglycerides, cerebrosides, phosphatide, phospholipids, terpenes, fatty acids, and fatty alcohol (Akoh and Min, 2008; Belitz et al., 2009; Chow, 2008). they're in natural sources like plants, animals, and insects. In recent years, the food business has centered on lipids to use them in edible films and coatings for preservation, however, edges for adding on food square measure vast. Akoh and Min (2008) represented however lipids in edible film and coatings offer several options, as an example, they minimize wet loss, offer gloss, cut back complexness, and value of packaging. The presentation of lipids might have an effect on some options in film or coating and food, the wet barrier being the foremost affected. Animal and plant waxes have a better potency barrier wet than resins, monoglycerides, diglycerides, fatty alcohols, emulsifiers, and surface active agents (Huber and Embuscado, 2009).

3.1 OILS AND FATS

Oils and fats square measure mixtures wherever the key compounds square measure triglycerides; they are available from plants and animals, severally. This mixture is with chemicals similar however differs physically, as oils square measure liquids and fats square measure solids (Igoe, 2011; Wool and Sun, 2005). Fig. 1.2 represents with chemicals the lipids compounds that square measure common on totally different oils, like vegetable oil, olive oil, palm oil, corn oil, vegetable oil, rice bran oil, vegetable oil, vegetable oil, palm oil; solely butter cocoa, vegetable oil, and vegetable oil square measure exempt of linolenic (Chow, 2008).

3.2 WAXES

Waxes have higher mass as a result of they're shaped by esters of an extended chain acid and long chain alcohol. The origin of waxes is animal and vegetal; they need a operate of protecting covering tissues. These square measure helpful on edible films or coatings for potency reducing wetness permeableness for top property (Akoh and Min, 2008; Huber and Embuscado, 2009; Sikorski and Kolakowska, 2011).

3.3 RESINS

Resins square measure substances that plant cells turn out for response to injury or infection in trees and shrubs; and a few insects will turn out them, that is that the case of Lacciferlacca that produces shellac organic compound. Major of resins square measure clear with yellowish-brown tones and physically square measure solid or solid (Baldwin et al., 2012).

3.4 PLASTIFICIZERS

Plasticizers square measure compounds with low mass that increase flexibility and strength of a fabric. The addition of plasticizers on film or coating facilitate increase permeableness to water and gases because of capability for reduction of unit forces in a very compound. glycerine and polysorbates square measure common plasticizers (Han, 2014; Rahman, 2007). The addition of various plasticizers lipidic to film or coating have a positive have an effect on , that is that the case glycerol-sage seed gum film, wherever incontestable plasticizers increase thickness, wetness content, wetness uptake (Mohammad et al., 2015), however mechanical properties (elongation and tensile strength) were influenced by concentration, hydrophobic tail of compound, and stirring of emulsion that was the case of films with glycerol-chia seed mucilage and glycerol-chitosan (Dick et al., 2015; Santacruz et al., 2015). Regarding morphology, glycerol-crees seed gum edible films were solid and sleek while not cracks, as delineate by Jouki et al. (2013).

3.5 EMULSIFIERS

Emulsifiers square measure molecule stabilizers of character ionic which will scale back physical phenomenon between 2 unmixable phases at their interface, permitting them to become compatible. The principal operate is preventing syneresis or phases separation, as a result of they keep hydrophilic–lipophilic balance (Badui, 2006; Igoe, 2011; Rahman, 2007). vital emulsifiers square measure lecithins; they're a mix or fractions of phospholipids, they originate in animal-like egg phospholipid or vegetal-like soybean lecithins (Whitehurst, 2004). Soy phospholipid has effects on edible films or coatings, as an example, color, opacity, solubility, and in microstructure. Fadini et al. (2013) connected opacity and color yellow of the coatings for soy phospholipid however Andreuccetti et al. (2011) associated action of phospholipid with solubility and microstructure. Andreuccetti et al. (2011) determined that phospholipid films, on study of research, had the very best concentration of phospholipid in film presented; tiny globules on the surface indicate heterogeneousness in supermolecule network.

4. PROTEIN-BASED EDIBLE FILMS

The edible packages proposal relies on victimisation biopolymers. The goal is to affix the principal characteristics and qualities of every one which method acquire the most effective result, which might be proteins, lipids, or polysaccharides (Quintero et al., 2010). Also, alternative objectives of the EF is to collect enough qualities therefore it will nourish the foodstuff and conjointly defend it from unhealthy organism by antimicrobial unharness (Lin and Zhao, 2007).

5. EDIBLE PACKAGING FOR FOOD APPLICATION

The quantity of packaging materials has been raised by 8 May 1945 annually (Muizniece-Brasava et al., 2011; Tavassoli-Kafrani et al., 2015). but five-hitter of all plastics square measure recycled, resulting in a high accumulation of plastics within the atmosphere (Espitia et al., 2014a,b). Continuous removal of those plastics (nonbiodegradable), to get a growing tendency to exchange petroleum-based polymers with perishable materials, these perishable polymers are used in the main to formulate edible films and coatings (De Azeredo et al., 2014). Edible films and coatings are assessed in numerous varieties of food, with glorious leads to prolonging the time period (Falguera et al., 2011). Here square measure some applications on fruits and vegetables, dairy farm merchandise, meat and poultry, and therefore the main results that square measure obtained.

5.1 FRUITS AND VEGETABLES

Fresh and minimally processed fruits and vegetables square measure extremely perishable (Jung and Zhao, 2016), in order that they would like best postharvest technologies so as to keep up their storage and stability. To prolong the food time period there square measure many varieties of biopackaging made of natural polymers, like starch (Chiumarelli and Hubinger, 2014), cellulose (Taqui and Stamatin, 2014), chitosan (Elsabee and Abdou, 2013), galactomannans (Cerqueira et al., 2011), alginate (Guerreiro et al., 2015), proteins (Belyamani et al., 2014), and wax (Ochoa et al., 2011). Edible films (F) and coatings (C) are thought-about a technology with nice potential to increase the time period of fruits and vegetables (Cerqueira et al., 2011). Emulsions have tried to be effective to extend stability of a spread of contemporary or processed fruits, vegetables, cheeses, and alternative food merchandise (Galus and Kadzi ska, 2015). In Table 1.11 we have a tendency to describe a number of the applications for edible packaging and conjointly the most elements used.

5.2 FRESH FRUIT AND VEGETABLES

In recent years, shoppers have shown a growing interest within the consumption of contemporary manufacture, healthy and wholesome merchandise that bring health edges, that the demand for freshcut fruits and vegetables is extremely high. However, the peeling and cutting operations could accelerate the metabolic activities of plant structure (Chiumarelli and Hubinger, 2012), creating the minimally processed product a lot of perishable than intact fruits and vegetables. an answer for increasing the time period of contemporary and minimally processed fruits and vegetables is coating with edible film (Table one.12) (Ciolacu et al., 2014).

5.3 DAIRY FARM MERCHANDISE

Milk and dairy farm merchandise square measure a vital supply of the nutrients needed for growth in kids and for maintaining health in adults (Cardador and Gallego, 2016). Cheese is that the most various cluster of dairy farm products; but, the time period of cheeses is restricted by the action of microorganisms. Antimicrobial coatings applied to the cheese surface square measure a method to limit negative changes, which might be promoted throughout handling and storage (Galus and Kadzi ska, 2015). Table 1.13 summarizes totally different applications of edible coating and films in dairy farm merchandise.

5.4 MEAT AND POULTRY

Meat is characterised as a food with tissue structure and is usually treated by totally different sorts of technological processes that favor the event of microorganisms. Antimicrobial and intelligent packaging has emerged as a food-safety hurdle technology in recent years. Also, considering giant volumes and styles of ready-to-eat meat and poultry trade there's would like for various strategies to manage food-borne pathogens and to enhance the time period in their merchandise (Guo et al., 2014). By this fashion, edible films and coatings are accustomed improve the gas and wet barriers, mechanical properties, and sensory perceptions of meat merchandise (Galus and Kadzi ska, 2015). the applying that has been used is by foaming, dipping, spraying, casting, brushing, individual wrapping, or rolling (Ustunol, 2009). In recent years there are generated numerous sorts of combined meat preservers victimisation bioactive edible films and coatings (Table one.14).

5.5 FISH

Fresh and minimally processed fish is very perishable throughout storage in the main because of speedy microbic growth of microorganisms naturally gift within the fish or from contamination, which might sometimes lead to either economic or health-related issues (Gómez-Estaca et al., 2010; López Diamond State Lacey et al., 2014). For conservation, some ways are recently developed. one amongst these ways is that the style and application of edible films and coatings to enhance the protection and time period of fish. many polysaccharides are used, like chitosan, agar, starch, whey supermolecule isolate. Also, alternative techniques square measure accustomed prolong the time period, like marinating (Van Haute et al., 2016) to make a further sensory price. A outline of a number of the advances in analysis throughout the last years is given in Table one.15.

6. REGULATORY ASPECTS AND COMMERCIALIZATION OF EDIBLE PACKAGING

6.1 EDIBLE PACKAGING LEGISLATION AND CONSUMER BEHAVIOR

Since the last decade of the Seventies the eu Union (EU) has established a restrictive policy for analysis and internal control of food packaging and every one food contact materials for client health protection and industrial reasons, which has steel, wood, aluminum, cellulose, paper and plastics, and active and intelligent materials (Silano and Rossi, 2015). within the case of the USA, the legislation is regulated by the Food and Drug Administration (FDA). But, why within the last twenty years has edible and active packaging become therefore important? Well, it's as a result of it's thought-about an important strategy of the particular effort to accomplish food safety round the world (Barone et al., 2015). The government agency establishes variations between a Food Contact Substance (FCS), Food Contact Material (FCM), and Food Contact Article (FCA).

FCS is outlined as any substance that's supposed to be used as a part of materials employed in producing, packing, packaging, transporting, or holding food, if such use of the substance isn't supposed to own any technical impact within the food. FCM may be a single substance, like a chemical compound or associate inhibitor in an exceedingly chemical compound. As a substance, it's moderately pure. The FCA is that the finished film, bottle, dough hook, tray, or no matter is created out of the FCM. The definitions are often a touch confusing; but, each encompasses a single however substantial distinction. to raised perceive these definitions we will say that associate FCM is created with Farm Credit System and alternative substances wherever the composition are often variable, typically it's a combination like associate inhibitor in an exceedingly chemical compound. The analysis of food active and/or intelligent packaging follows identical principles for any FCM, relating to exposure assumptions of materiamedica knowledge. yet, the particular info regarding the functions of the substance or active ingredients within the packaging should be provided. in addition, migration testing is also effectuated with the particular conditions specially elite and designed for the supposed use of the packaging.

6.2 ECONOMIC FEASIBILITY AND BENEFITS OF EDIBLE PACKAGING

One of the foremost necessary parameters to require under consideration (if not the foremost important) throughout development and development of any product is that the economic practicability, as a result of despite if the merchandise or during this specific case edible films area unit glorious alternatives to ancient packaging if it's not economically profitable and technically potential. For that, the challenge to beat for edible coatings is that the low value of fossil materials, from that ancient plastics area unit made; throughout the last year's plastic from fossil sources has been a budget thanks to package food, not just for FCM, however additionally the baggage wherever food is transported from store to shopper homes. Notwithstanding, in recent years economic sources of biopolymers are explored to form edible films economically competitive with ancient packaging, that has resulted really in very viable choices. Van Beilen and Poirier (2007) attributes the rise of investigation of biopolymers for shopper and industrial functions to a few causes: (1) economic science, (2) public acceptance, and (3) laws.

Plastic waste represents a large downside round the world. utilisation isn't one resolution, as really but five-hitter of plastic waste is recycled (Espintia et al., 2014) and therefore the different ninety fifth remains as waste within the atmosphere. shoppers settle for that edible films will facilitate to decrease this downside. Among the advantages of edible packagings, they're perishable, while not toxicity, they'll scale back wet loss, and unfavorable reaction within the food, like accelerator browning (Osorio et al., 2011), spoilage, and microbic development (Arvanitoyannis, 2010) by adding bioactive compounds.

6.3 DEVELOPMENT ASPECTS OF THIS SORT OF PACKAGING

One of the primary business interests associated with edible films could be a patent submitted by the FMC Corporation, with the intention to use a wax coating in recent fruit (De Ore, 1933). Another formulation was given within the same year by gryphon Laboratories, exploitation gelatin coating for food preservation; foods which will be lined with this film embrace meat, fruits, prepackaged food, and vegetables (Griffith and Hall, 1933). Since then, however particularly within the last thirty five years, development of edible films and coatings has swollen as detected by Scourge of the Gods E. Pavlath (Bernstein and Woods, 2013) UN agency says that within the middle of Nineteen Eighties solely ten corporations were within the business of edible packaging, compared with quite one thousand that truly area unit currently which have Associate in Nursing annual financial gain for nearly $100 million (USD). Current applications of edible films and coatings area unit expected to still grow, as an example, within the use of ready-to-eat product that truly represent 100% of the entire manufacture sales.

In existence, patent requests for inventions as edible films Associate in Nursingd coatings for food applications enclosed Associate in Nursing ultrathin edible coating made up of gold or silver in an edible substrate, claimed by Ds cluster (Kumar and Kumar, 2015), the employment of algae, paper fibers, and starch (Santos, 2013), and a movie from carbohydrate gum as pullulan replacement (Nieto, 2012). Table 1.16 summarizes some corporations as samples of this commercially out there edible food packages. one in all the foremost necessary concerns to require under consideration to commercialize edible, natural, perishable, or active packaging for food application is that the current legislation of the country wherever the merchandise is to be oversubscribed. With the aim to possess a legal product, all of the ingredients contained within the edible package shall be approved to be used in food, as ingredients, additives, or as FCM, relying of the country.

REFERENCES

- Aguirre-Joya, J.A., De Leon-Zapata, M.A., Alvarez-Perez, O.B., Torres-León, C., Nieto-Oropeza, D.E., Ventura-Sobrevilla, J.M., Aguilar, M.A., Ruelas-Chacón, X., Rojas, R., Ramos-Aguiñaga, M.E. and Aguilar, C.N., 2018. Basic and applied concepts of edible packaging for foods. In Food packaging and preservation, Academic Press, pp. 1-61.
- Ahmad, V., Khan, M.S., Jamal, Q.M.S., Alzohairy, M.A., Al Karaawi, M.A. and Siddiqui, M.U., 2017. Antimicrobial potential of bacteriocins: in therapy, agriculture and food preservation. *Int. J. Antimicrob. Agents.*, 49, 1-11.
- Ahvenainen, R., 2003. Active and intelligent packaging: an introduction. In Novel food packaging techniques. pp. 5-21.
- Ali, A., Maqbool, M., Ramachandran, S. and Alderson, P.G., 2010.Gum arabic as a novel edible coating for enhancing shelf-life and improving postharvest quality of tomato (*Solanumlycopersicum* L.) fruit. Postharvest Biol. Technol., 58, 42-47.
- Badr, K.R., Ahmed, Z.S., El Gamal, M.S., 2014. Evaluation of the antimicrobial action of whey protein edible films incorporated with cinnamon, cumin and thyme against spoilage flora of fresh beef. *Int. J. Agr. Res.*9, 242.

- Badui, S., 2006.Química de los Alimentos, fourth ed. Pearson Educación, Naucalpan de Juárez.
- Bosquez-Molina, E., Guerrero-Legarreta, I. and Vernon-Carter, E.J., 2003. Moisture barrier properties and morphology of mesquite gum–candelilla wax based edible emulsion coatings. *Food Res. Int.,* 36, 885-893.
- Bourtoom, T., 2008. Edible films and coatings: characteristics and properties. Int. Food Res. J., 15, 237-248.
- Coles, R., McDowell, D. and Kirwan, M.J. eds., 2003.Food packaging technology (Vol. 5). CRC Press.
- Cuero, R.G., Osuji, G. and Washington, A., 1991. N-carboxymethylchitosan inhibition of aflatoxin production: role of zinc. Biotechnol.Lett., 13, 441-444.
- Dangaran, K., Tomasula, P.M. and Qi, P., 2009.Structure and function of protein-based edible films and coatings.In Edible films and coatings for food applications. Springer, New York, NY, pp. 25-56.
- Donhowe, I.G. and Fennema, O., 1993. The effects of plasticizers on crystallinity, permeability, and mechanical properties of methylcellulose films. J. FoodProcess. Preserv., 17, 247-257.
- Dragich, A.M. and Krochta, J.M., 2010. Whey protein solution coating for fat-uptake reduction in deep-fried chicken breast strips. J. Food Sci., 75, S43-S47.
- Embuscado, M.E. and Huber, K.C., 2009.Edible films and coatings for food applications (Vol. 9). New York: Springer.
- FAO, 2011. Global Food, and Food Waste. "Extent, causes and prevention." Rome: Food and Agriculture Organization of the United Nations.
- Gómez-Estaca, J., López-de-Dicastillo, C., Hernández-Muñoz, P., Catalá, R. and Gavara, R., 2014.Advances in antioxidant active food packaging. Trends Food Sci. Technol., 35, 42-51.
- Gontard, N., Duchez, C., CUQ, J.L. and Guilbert, S., 1994. Edible composite films of wheat gluten and lipids: water vapour permeability and other physical properties. International journal of food science & technology, 29, 39-50.
- Greener, I.K. and Fennema, O., 1989.Barrier properties and surface characteristics of edible, bilayer films. J. Food Sci., 54, 1393-1399.
- Guilbert, S., Gontard, N. and Cuq, B., 1995.Technology and applications of edible protective films. Packaging Technology and Science, 8, 339-346.
- Hadian, M., Rajaei, A., Mohsenifar, A. and Tabatabaei, M., 2017. Encapsulation of Rosmarinusofficinalis essential oils in chitosan-benzoic acid nanogel with enhanced antibacterial activity in beef cutlet against Salmonella typhimurium during refrigerated storage. LWT, 84, 394-401.

- Hambleton, A., Debeaufort, F., Bonnotte, A. and Voilley, A., 2009. Influence of alginate emulsion-based films structure on its barrier properties and on the protection of microencapsulated aromacompound. Food Hydrocoll., 23, 2116-2124.
- Han, J.H., 2014. Innovations in food packaging, Elsevier, Academic Press, USA.
- Hannon, J.C., Kerry, J.P., Cruz-Romero, M., Azlin-Hasim, S., Morris, M. and Cummins, E., 2017. Kinetic desorption models for the release of nanosilver from an experimental nanosilver coating on polystyrene food packaging. Innov.Food Sci. Emerg. Technol., 44, 149-158.
- Janjarasskul, T., Rauch, D.J., McCarthy, K.L. and Krochta, J.M., 2014. Barrier and tensile properties of whey protein–candelilla wax film/sheet. LWT- Food Sci. Technol., 56, 377-382.
- Kester, J.J. and Fennema, O.R., 1986. Edible films and coatings: a review. Food Technol. (USA).
- Krochta JM, De Mulder-Johnston C., 1997. Edible and biodegradable polymer films: challenges and opportunities. Food Technol., 51, 61–73
- Lin, L., Abdel-Shafi Abdel-Samie, M., & Cui, H., 2018. Novel Packaging Systems in Food.Reference Module in Food Science.484-491.
- MacGowan, A., Macnaughton, E., 2017. Antibiotic resistance. Med. Baltim. 45, 622–628.
- Marsh, K. and Bugusu, B., 2007. Food packaging—roles, materials, and environmental issues. J. Food Sci., 72, 39-55.
- Nordström, R. and Malmsten, M., 2017.Delivery systems for antimicrobial peptides. Adv. Colloid Interface Sci., 242, 17-34.
- Oussalah, M., Caillet, S., Salmiéri, S., Saucier, L. and Lacroix, M., 2004. Antimicrobial and antioxidant effects ofmilk protein-based film containing essential oils for the preservation of whole beef muscle. J. Agr. Food Chem., 52, 5598-5605.
- ParenteRibeiroCerqueira, M. Â., 2018. Edible Packaging.Reference Module in Food Science.173-176.
- Ribeiro-Santos, R., Andrade, M. and Sanches-Silva, A., 2017.Application of encapsulated essential oils as antimicrobial agents in food packaging.Curr. Opin. Food Sci., 14, 78-84.
- Rodrigues, S. and Fernandes, F.A.N., 2012. Advances in fruit processing technologies.CRC Press.

- Suppakul, P., Miltz, J., Sonneveld, K. and Bigger, S.W., 2003.Active packaging technologies with an emphasis on antimicrobial packaging and its applications. J. Food Sci., 68, 408-420.
- Šuput, D.Z., Lazi , V.L., Popovi , S.Z. and Hromiš, N.M., 2015. Edible films and coatings: Sources, properties and application. Food and Feed Research, 42, 11-22.
- Thomas, L.V. and Delves-Broughton, J., 2005.7 Nisin. Antimicrobials in food, 237.
- Trinetta, V., 2016, "Biodegradable Packaging." Reference Module in Food Science, 2-4.
- Wittaya, T., 2012. Protein-based edible films: Characteristics and improvement of properties. Structure and function of food engineering, 43-70.

7

CHAPTER

RECENT TRENDS IN ACTIVE, SMART AND INTELLIGENT PACKAGING FOR FOOD PRODUCTS

Nadira Anjum*[1], Anju Bhat[1], Fozia Hameed[1], Anjum Ayoub[1], Mohd. Aaqib Sheikh[2] and Rukhsana Rahman[1]

[1]*Division of Food Science and Technology, Sher-e-Kashmir University of Agricultural Sciences and Technology- Jammu, Chatha-180009, India*

[2]*Department of Food Engineering and Technology, Sant Longowal Institute of Engineering and Technology, Longowal, Punjab-148106, India*

Corresponding Author Email ID: nadiraanjum1736@gmail.com

ABSTRACT

The changes in consumer preferences for safer as well as hygienic food have lead to many innovations in the technologies of packaging and traditional packaging is considered as no longer sufficient to meet the consumer demands. In the life cycle of food product, the role of packaging cannot be neglected. Active and intelligent packaging involves a deliberate interaction between the packaging and the food product or the environment surrounding the product. The integration of active and intelligent materials into the packaging systems is expected to have a promising future. These newer techniques work synergistically to yield a packaging system with multiple benefits. The importance of these techniques is proven for maintaining food quality and safety. These techniques need full emergence in research and market because the benefits of these opportunities is contrasted by a number of challenges that need to be overcome to reach the full potential.

Keywords: Packaging, safety, food, hygiene, consumer

1. INTRODUCTION

Traditional packaging of food is considered to provide protection, communication, containment and convenience. Food packaging provides protection to the food product from the external surrounding environment of the product. Packing of food saves the time and makes it easier for the consumers to handle the packages. Inertness of the packages in direct contact with the food materials is an important safety concern i.e, minimum interaction between the traditional packaging materials and the food is considered as an important criterian in selecting the proper packaging materials (Marsh and Bugusu, 2007). On the other hand, smart packaging techniques like active and intelligent packaging involve a useful interaction between the food product and the packaging material which results in protection of the food material. New packaging techniques are usually described by the terms like 'active', 'interactive', 'smart', 'clever', 'indicators' etc. Global trends like industrial processing of food at an increased rate, requirement of less time for preparation of food products, import and export of food products has lead to the need of development of such packaging materials that provide healthier, safer, convenient and cheaper food products. Extension of shelf life by applying a proper packaging material results in the prevention of losses of food products. The consumers demand for the safe food acts as a driving force for the development of newer as well as improved technologies for packaging as a result the packaging industries majorly focus on the packaging materials with improved gas barrier properties, transparency, UV protection and environmental performance etc (Dainelli, *et al.*, 2008). Other major factors that lead to the development of smart packaging systems are traceability of a particular product, indication of temper and sustainability. Active and intelligent packaging techniques are considered as an innovation to the packaging systems and the global market for such newer packaging techniques is estimated to double between the years 2011 and 2021. Experts consider these techniques as the future of packaging. Advanced packaging contributes about 5% to the total packaging market value out of which smart packaging systems make about 11% (Pereira *et al.*, 2012).

The innovations to the packaging materials initially were limited to a small number of materials used as a barrier like complex and multilayer materials, new polymers etc. The first series of newer packaging techniques in developed countries were the use of moisture and oxygen scavengers for successfully increasing the shelf life and improving the quality of food products like meat. Next to that various other active and intelligent packaging systems were developed like ethanol emitters for bakery products, ethylene absorbers for climacteric fruits, time temperature indicators, oxygen indicators etc. in general we can say that the field has been extended to increase the market opportunities by providing a series of niche markets. The newer techniques of packaging provide a number of innovative ways for improving the shelf life as well as maintaining the quality of various food products (Rooney, 2005). Sensors used in the intelligent packaging help to detect the microbes as well as contaminants and hence help in maintenance of food safety. It actually gives information about the condition of the food product to give idea about its quality, safety, shelf life as well as the usability of the product. In other words the newer techniques make use of a combination of smart as well as conventional packaging materials which result in improved benefits of the packaging to the food products. The smart materials used should be accurate, reliable, simple as well as reliable (Kerry, et al., 2006). There are various indications that the sales for intelligent packaging systems will grow rapidly in the next coming years as the number of patent applications and the number of patents granted for such packaging systems is increasing these years. The concepts of modern packaging techniques were first introduced in USA and Japan. Nowadays Europe is introducing such technologies in the field of packaging despite the restrictive legal regulations and the number of research projects on development of such technologies in Europe is also increasing (Vanderroost, *et al*. 2014).

The deliberate interaction of the active and intelligent packaging systems with the food or the environment surrounding in comparison with the traditional packaging systems leads to the challenges for evaluation of its safety as the migration of some substances from packaging to the food is considered as a major risk. Main and important objective of the food law is to ensure food safety. Quality control in manufacturing of food products is related to its nutritive value, physical as well as sensory attributes, microbiological safety and the chemical properties of the product (Lim,*et al*.,2014). Other risks may be related the incorrect use or an inefficient use of packaging system. Hence the provisions related to the safety of newer packaging techniques like active and intelligent packaging which are intended to come in close contact with the food materials are included in the Regulation 1935/2004 on materials and articles in close contact with food. In other words smart packaging technologies in context of globalization serve as a means of increasing the market value but at the same time results in accommodation of stricter food safety regulations at both national as well as international level that prevents the chances of occurance of food bioterrorism (Yam, *et al*., 2012).

2. FUNCTIONS OF PRODUCT PACKAGING

First and the most important function of packaging is to protect the food from the deteriorative changes caused by the environment surrounding the package. In addition to protection function, packaging of a products acts as an effective tool of marketing by communicating with the consumers. Consumers are provided with an ease of use as well as convenience by packaging a product (Robertson, 1993). The main functions of packaging are:

- By packaging, a product is prevented from any type of leak or breakage as well as from external contamination.
- With the help of packaging, the useful information related to the nutritive value, cooking properties etc related to the contained food product is communicated to the consumers of that particular product. This function of packaging is regulated by law and the proper display of the information related to the food product will have an influence on the consumer acceptance of the particular product. The product can be available in different shapes and sizes by providing a proper packaging.
- Packaging also provides convenience to the consumers e.g. it becomes easy to reheat the packaged food product in the microwave.

- Packaging of the product provides containment, which results in ease during handling as well as transportation of the package. Sales as well as consumer attitudes of a particular food product is affected by the expectations like improved physical characteristics and containment aspects of the packaging (Tu, *et al.*, 2015).

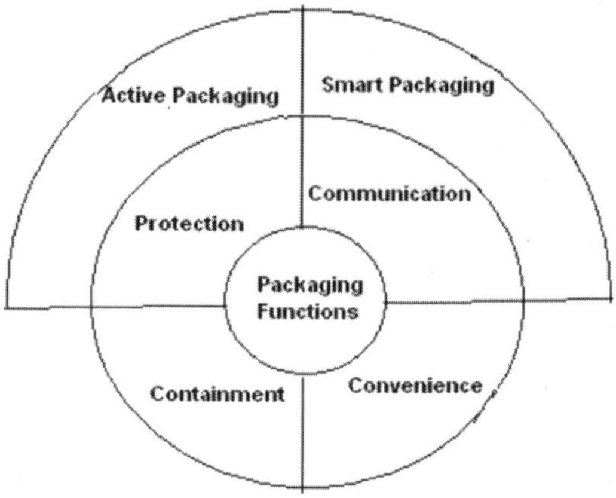

Fig 1: Model for packaging functions

3. ACTIVE PACKAGING:

Active packaging is considered as a substitute to the traditional packaging methods. In active packaging, active components are embedded into the packaging that absorb or release certain substances from or into the food product or the environment surrounding it in order to prolong shelf life and maintain food quality. Oxygen scavengers, ethylene scavengers, flavour and odour absorber/releaser, antimicrobial and antioxidants are the components that are most commonly used in active packaging. It is an alternative to the use of preservatives as well as to the modified atmosphere packaging. Active packaging provides a different role other than providing a non reactive and inert barrier between a food product and the environment surrounding it. In active packaging a positive interaction occurs between the product and the package in order to improve the quality as well as the acceptability of the food (Miltz *et al.*, 1995). Active packaging plays very important roles like controlling the process of respiration in various horticultural commodities, preventing or delaying oxidation in muscle foods, preventing moisture migration and growth of microbes in dried foods. The gas concentration inside the package is modified to the levels which are necessary for increasing the shelf life by manipulating the permeability of packaging materials (Prasad and Kochhar, 2014). The various components used in active packaging are:

3.1 OXYGEN SCAVENGERS:

Oxygen scavengers are most widely used among the active ingredients used in the packaging. Oxygen scavengers prevent the deteriorative effects of oxygen present in the package by scavenging it. By the use of oxygen scavengers the oxygen levels in the headspace of package is reduced to less than 100 ppm. The scavengers available commercially are based on one or more of these technologies iron powder oxidation, enzyme oxidation, ascorbic acid oxidation, saturated fatty acid oxidation, photosensitive dye oxidation, immobilized yeast on solid material etc. (Floros, *et al.*, 1997).

3.2 CARBON DIOXIDE ABSORBERS AND EMITTERS

Carbon dioxide is incorporated in the food package in order to prevent microbial growth in products like fish, meat, poultry as well as bakery products. It is also desirable to add carbon dioxide in packages with fresh foods as it will decrease the respiration rates. Carbon dioxide scavengers are used to remove the excess amount of CO_2 from the packages containing products like roasted coffee sealed hermatically and the accumulation of CO_2 inside it may lead to bursting of the package.

3.3 ETHYLENE SCAVENGERS:

Ethylene is a natural plant growth harmone that is responsible for accelerating the rate of respiration in fresh fruits and vegetable which inturn causes fruit ripening, softening and senscence at very low levels (Abeles *et al.*, 1992). The most commonly used ethylene scavenger is potassium permaganate immobilized on inert minerals and is available in the form satchets. Other ethylene scavengers used are activated carbon base with various metalscatalysts, activated charcoal impregnated with palladium catalyst, activated clay embedded in plastic bags (zeolite) (Day 2008).

3.4 ANTIOXIDANT RELEASE:

Antioxidants improve the oxidative stability of the food products. In active packaging, antioxidants are incorporated into the packaging material which is released at the time of requirement. The antioxidants incorporated into the packaging can prevent the polymers from degradation. The most common antioxidants incorporated in the package are BHT, vitamin C and E etc. These antioxidants have proven to be stable under processing conditions (Vermeiren, *et al.*, 1999).

3.5 MOISTURE CONTROL:

We know that the presence of moisture inside the package can lead to the spoilage of the product. Hence the control of excess moisture in the package should be applied. Moisture absorbing pads, blankets and sheets are used for this purpose in food products like meat, fish, poultry, fruits and vegetables. Also dessicants like silica gel can be used to prevent moisture accumulation inside the packages containing dry foods.

3.6 ANTIMICROBIAL PACKAGING:

Antimicrobial packaging is a type of active packaging that plays the role of reducing, preventing or retarding the growth of microbes present inside the package or on the package itself in order to maintain the quality as well as safety of the food. In this type of active packaging the antimicrobial agents are either incorporated into the packaging material or coated on it. Natural antioxidants may be either the extracts obtained from spices like clove, cinnamon, rosemary etc or derived from the substances produced from the bacterial or fungal action like the polypeptide nisin, natamycin (Han, 2000).

Other active packaging systems include flavor and odor absorbers and releasers, self heating and self cooling, self venting and microwaveable active packaging etc. microwavable active packaging involves the use of microwave susceptors. Microwave susceptors are meant for the even heating of a product as it consists of stainless steel or aluminum particles deposited on polyester or paper board like substrates. Microwave susceptors also cause crispiness and surface browning of products which otherwise is a drawback of microwave heating or cooking (Vermeiren, *et al.*, 1999).

Table 1: Commercially available active packaging systems

Trade Name	Principle	Type
Ageless	Iron based	Oxygen scavenger
Freshilizer	Iron based	Oxygen scavenger
Freshmax, Freshpax, Fresh Pack	Iron based	Oxygen scavenger
Oxyguard	Iron based	Oxygen scavenger
Zero$_2$	Photosensitive dye	Oxygen scavenger
Bioka	Enzyme base	Oxygen scavenger
Dri-Loc®	Absorbent pad	Moisture absorber
Tenderpac®	Dual compartment system	Moisture absorber
Neupalon	Activated carbon	Ethylene scavenger
Peakfresh	Activated clay	Ethylene scavenger
Evert-Fresh	Activated zeolites	Ethylene scavenger
SANICO®	Antifungal coating	Interleavers
Agion®	Silver based	Antimicrobial packing
Biomaster®	Silver based	Antimicrobial packing

Source: Biji *et al.*, 2015

4. INTELLIGENT PACKAGING:

Intelligent packaging systems give an idea about the conditions of the food product as well as the environment surrounding it. It is considered as an extension of the communicative functions of the packaging. Intelligent materials can be defined as the materials that are used to monitor the condition of product inside the package as well as the environment surrounding it. Intelligent packaging provides information to the consumers based on its ability to sense, record and detect the changes in the package as well as in the surrounding environment (Realini and Marcos, 2014). The most commonly used components of intelligent packaging are immobilized enzymes as well as antibiotics. Intelligent packaging employs the use of various types of indicators like time temperature indicators, immobilized bioactive components, chemical, spoilage and various types of contamination indicators. This type of packaging is also known as smart packaging and is different from smart packaging. It is also used to check the efficacy as well as reliability of active packaging systems.

4.1 TYPES OF INDICATORS IN INTELLIGENT PACKAGING:

The three types of devices used in the intelligent packaging are given below:

4.1.1 External indicators: these are the indicators placed at outside of the package. It gives an indication about the surrounding conditions of the package. It includes time, temperature and physical shock indicators.

4.1.2 Internal indicators: these types of indicators are either placed in the headspace of the package or are attached to the inner side of the lid of package. It includes microbial indicators, oxygen leak indicators etc.

4.1.3 Indicators that increase the efficacy of flow of information. These employ special type of barcodes that store information related to the product and hence results in an effective communication between the product and the user. It gives information about the product consumption procedure, expiry and manufacture date of product, storage conditions.

Table 2: Example of external and internal indicators and their principles

Techniques	Type	Principles/Reagents	Information Provided	Applications
Time temperature indicators	External	Mechanical/chemical/ enzymatic reaction	Storage conditions	Foods stored under chilled and frozen conditions
Carbon dioxide indicators	Internal	Chemical reactions	Storage condition and package leak	Modified or controlled atmosphere food packaging
Oxygen indicators	Internal	Redox dyes, pH dyes, enzymes	Storage conditions package lack	Foods stored in packages with reduced oxygen concentration
Pathogen indicator	Internal	Various chemical and immunochemical methods reacting with toxins	Specific pathogenic bacteria, such as *E. coli* O:157	Perishable foods, such as meat, fish, and poultry.
Microbial growth indicators and freshness indicators	Internal/ external	Ph dyes, all dyes reacting with certain metabolites	Microbial quality of food (i.e., spoilage)	Perishable foods, such as meat, fish, and poultry

Source: Ahvenainen and Hurme, 1997

5. AREAS OF APPLICATION

It has been found that in United States number of cases of food borne diseases per year are more than 6 million resulting in more than 90,000 deaths per year. In Spain, there occur about 60 cases per 10,000 inhabitants per year (Pereira, *et al., 2012).* So the most important application of newer packaging techniques is the development of sensors for detection of pathogens in food. Apart from the sensors, the emitters, absorbers and scavengers embedded in the packaging as active packaging materials also find wide range of applications in monitoring the food safety and maintenance of quality. The smart packaging techniques can offer a number of value added benefits like drug use, delivery tracking of the items with embedded security tags (Arvanitoyannis and Stratakos, 2012). Use of smart packaging techniques helps to identify the inefficiencies in supply chains of the products. These techniques also help in increasing the profit margins by reducing the costs as well as errors and increasing the product performance (Pacquit et al., 2007).

6. RESEARCH CHALLENGES AND OPPORTUNITIES

It has been studied that about one third of the edible food products meant for human consumption are wasted due to inefficient and inappropriate practices during harvesting, handling, transportation and storage of such edible commodities. This huge waste of food is a financial threat to the food industries and hence such practices should be designed that prevent such losses of food (Morris, et al., 2017). One among such practices includes design and development of smart packaging techniques. The challenges and opportunities for the development of such techniques are given below:

6.1 CHALLENGES

The various challenges of new packaging techniques are given below:

- Integration of thin film electronics into the packaging material for monitoring the temperature for prediction of shelf life of food products has some issues related to the performance and thus research and development in this area is considered as a challenge.
- We know that active packaging helps to improve safety and shelf life of food but nowadays there is a growing need for the use of active agents obtained from natural sources and the biodegradable polymeric materials as carriers of these active agents (Realini and Macros, 2014).
- The waste obtained from the smart packaging is considered to be non recyclable and that may lead to many problems to the packaging industries. So, there is a challenge to conduct research on the disposal of waste resulting from newer and smart packaging techniques and on the implementation of biodegradable materials that will allow integration of sensors with affecting its communication functions (Cruz, et al., 2012).

- Consumers expect more and more information about the food product to be conveyed to them like information about the ingredients, origin, handling, cooking and storage conditions etc. Hence the information embedded in the packaging in particular is a challenge to the food industries.
- Growing interest of researchers as well as food industries towards the use of antimicrobial packaging is due to the benefits like quality and safety of food. Focus of research should be towards the use of naturally obtained antimicrobial agents, bio-preservatives and biodegradable packaging materials. It has been seen that biodegradable packaging has already resulted in a number of innovations in the sector of packaging and thus contributed towards safety and quality of food (Siripatrawan and Noipha, 2012).

6.2 OPPORTUNITIES

The opportunities related to the development of modern packaging techniques are summarized below:

- Industrial internet of things helps in interconnection of the objects with the cyber world globally by providing a network. By this we can track the objects equipped with sensors e.g. the packaging containing a RFID tag can easily be tracked throughout the transportation and handling process.
- Nanotechnology is considered to play a very important role in near future for the development of appropriate, safe and advanced packaging materials designed for the controlled release of active ingredients.
- Cyber security is the main concern related to the smart packaging techniques. Due to the modern internet security, data privacy can be an issue that will lead to a negative effect on the potential of smart packaging techniques. In order to obtain defence in depth and robust cyber security systems such methodologies that ensure automatic detection as well as a response to cyber attacks should be employed synergistically with the prevention techniques.
- The implementation of smart packaging techniques leads to the development of new business models that can result in a number of opportunities. A shift can be seen in the industries from products to product service systems.

7. CONCLUSION

In recent years various types of active and intelligent packaging techniques are developing to be integrated into the packaging for the purpose of increasing the shelf life and maintaining the quality of food products. The advantages of these packaging techniques are quite obvious but have some related issues that need to be solved before the widespread applications of these types of techniques. Further research on the use of these newer food packaging techniques can improve the present systems and can offer opportunities which result in consumer benefits and convenience. The research will also eliminate the existing gap of these systems in marketing applications. Also it is important to make the consumers aware about the extra benefits of these packaging systems by providing them knowledge about these systems.

REFRENCES

- Abeles, F.B., Morgan, P.W. and Saltveit, M.E. 1992. Ethylene in plant biology. Academic press Inc, San Diego, California
- Ahvenainen, R. and Hurme, E.1997. Active and smart packaging for meeting consumer demands for quality and safety. *Food Additives and Contaminants,* **14**: 753–763.
- Arvanitoyannis, I.S. and Stratakos, A.C. 2012. Application of modified atmosphere packaging and active/smart technologies to red meat and poultry: a review. *Food and Bioprocess Technology,* **5(5): 1423-1446.**
- **Cruz, N.F., Simões, P., Arques. R. 2012. Economic cost recovery in the recycling of packaging waste: the case of Portugal.** *Journal of Cleaner Production,* **37: 8-18.**
- Dainelli, D., Gontard, N., Spyropoulos, D., Zondervan-van den Beuken, E. and Tobback, P. 2008. Active and intelligent food packaging: legal aspects and safety concerns. *Trends in Food Science and Technology,* **19**: 103–112.
- Day, B.P.F. 2008. Active packaging of food. In: Kerry J, Butler PJ (eds) Smart packaging technologies for fast moving consumer goods. Wiley and sons Ltd, West Sussex, England.
- Floros, J.D., Dock, L.L. and Han, J.H. 1997. Active packaging technologies and applications. *Food Cosmetics and Drug Packaging,* **20**:10–17.
- Han, J.H. 2000. Antimicrobial Food Packaging. *Food Technology,* **54**(3):56–65.
- Kerry, J.P., O'Grady, M.N.and Hogan. S.A.2006. Past, current and potential utilization of active and intelligent packaging systems for meat and muscle-based products: a review. *Meat Science,* **74**: 113–130.
- Marsh, K. and Bugusu. B. 2007. Food packaging: Roles, materials, and environmental issues. *Journal of Food Sciences,* **72**:39–55.
- Prasad, P. and Kochhar, A. 2014. Active packaging in food industry: a review. *Journal of Environmental Science, Toxicology and Food Technology,* **8(5): 1-7.**
- Pereira de Abreu, D.A., Cruz, J.M. and Paseiro Losada, P.2012. Active and intelligent packaging for the food industry. *Food Reviews International*, **28**: 146–187.
- Pacquit, A., Frisby, J., Diamond, D., Lau, K.T., Farrell, A., Quilty, B. and Diamond, D. 2007. Development of a smart packaging for the monitoring of fish spoilage. *Food Chemistry,* **102(2): 466-470.**

- **Realini, C.E. and Marcos, B. 2014. Active and intelligent packaging systems for a modern society.** *Meat science,* **98(3): 404-419.**
- **Siripatrawan, U. and Noipha, S. 2012. Active film from chitosan incorporating green tea extract for shelf life extension of pork sausages.** *Food Hydrocolloids,* **27(1): 102-108.**
- Lim, S.A.H, Antony, J. and Albliwi, S.2014. Statistical Process Control (SPC) in the food industry—a systematic review and future research agenda," *Trends in Food Science and Technology,* **37**(2):137–151.
- Miltz, J., Passy, N. and Mannheim, C.H.1995. Trends and applications of active packaging systems. In: Jagerstad M, Ohlsson M (eds) AckermanP. Food and packaging materials-Chemical interaction. The Royal Society of Chemistry, London, England, pp. 201–210.
- Morris, M.A., Padmanabhan, S.C., Cruz-Romero, M.C., Cummins, E. and Kerry, J.P. 2017. Development of active, nanoparticle antimicrobial technologies for muscle-based packaging applications. Meat Science, **132: 163-178.**
- Realini, C.E. and Marcos, B. 2014. Active and intelligent packaging systems for a modern society. *Meat Science*, **98**:404–419.
- Robertson, G.L. 1993. Food packaging: principles and practice. Marcel Dekker, New York. Pages 686.
- Rooney, M. 2005. Introduction to active food packaging technologies. In J. H. Han (Ed.), Innovations in food packaging. Elsevier Academic Press.
- Tu, Y., Yang,Z. and Ma, C. 2015.Touching tastes: the haptic perception transfer of liquid food packaging materials," *Food Quality and Preference*, **39**:124–130.
- Vanderroost, M., Ragaert, P., Devlieghere, F. and De Meulenaer, B.2014. Intelligent food packaging: The next generation. *Trends in Food Science and Technology,* **39**: 47–62.
- Vermeiren, L., Devlieghere, F., Van Beest, M., de Kruijf, N. and Debevere, J. 1999. Developments in the active packaging of foods. *Trends Food Science and Technology,* **10**:77–86.
- Yam, K.L., Takhistov, P.T. and Miltz, J. 2005. Intelligent packaging: concepts and applications. *Journal of food science,* **70: 1-10.**
- Biji, K.B., Ravishankar, C.N., Mohan,C.O and Srinivasa Gopal, T.K. 2015 Smart packaging systems for food applications: a review, *Journal of Food Science and Technology,* **52**(10):6125-6135.

8

CHAPTER

FOOD FREEZING TECHNOLOGIES

Anjum Ayoub*[1], Kundan Veer Singh[2], Monika Sood[1], Fozia Hameed[1] & Nadira Anjum[1]

[1]*Division of Food Science and Technology, Sher-e-Kashmir University of Agricultural sciences and Technoloy-Jammu, Chatha-180009, India*

[1]*Division of Plant Breeding and Genetics, Sher-e-Kashmir University of Agricultural sciences and Technoloy-Jammu, Chatha-180009, India*

*Corresponding Author: *[1]anjumparay77@gmail.com@gmail.com*

ABSTRACT

Freezing is one of the most commonly used and reliable methods for extending the shelf life of foods for extended periods of time. This chapter focuses on freezing at very low temperatures, and a thermal model was developed based on literature formulations to measure the food freezing period for a variety of foods of various sizes, shapes, and chemical compositions. Furthermore, various types of freezers and food freezing types are explained with their advantages. These various kinds of freezers help to preserve the food for longer periods of time without reducing the quality of the food products for longer periods.

Keywords: Low temperature freezing; freezers, Preservation, quality.

INTRODUCTION

Freezing is an unshakable food preservation process that provides high quality nutritious foods for long self-life. (Fikiin K, 2003).Freezing is the natural extended refrigeration process which was commercialized by Birdseye in 1928 and swiftly used to preserve a broad variety of foodstuffs. Each year, a massive quantity of foodstuff is produced to satisfy the request of patrons. Currently, legislations pertain to foodstuff storing method are becoming more deterring in terms of final quality of food products and of transformation and storage processes efficiency. (Biglia *et al.*, 2015). Fresh plant-based foods may not available all year round for consumption and the long-term storage of fresh foods could be provocation due to high water content, unavailability of cold-storage facilities (particularly in underdeveloped and developing countries), and possibility of nutritional deterioration, drying of such foods may allow their long-term consumption and eases handling, transportation, and storage. Traditional industrial food freezers are based on air-blast system technology (Dempsey P and Bansal P, 2012).Despite the novel technique enhancement that accompanied the industrial revolution, most preservation of foods still relies on the principles discovered by ancient cultures: retarding microbial population growth by using low temperature, dehydration, wood smoke, unbalanced nutritional composition, osmotic stress, or organic chemicals. Modern preservation techniques often combine multiple methods (e.g., added chemicals and freeze-drying) to increase the shelf life of industrially prepared foods. Even with the most modern techniques, the majority of the caloric requirements in contemporary industrial societies are typically met by cereal grains, which are still preserved primarily by simply keeping them dry (Archer, 2004).The huge lacuna during freezing though is the growth of ice crystals inside the foodstuff, which can damage internal structures.Around 80-90% of water is present in fruits and vegetables and due to freezing process, this liquid water converts to a solid form. These solid ice crystals are the chief factor disturbing quality in frozen foods, according to Professor Sun, because they can harm the internal structure of the food, such that when the food thaws out, the nutrients drip out of the food along with the water: "So we want to freeze the food without the internal damage, and this will preserve the nutrients inside the food. Currently, the food industry utilizes a blast of cold air or extremely cold liquid nitrogen or carbon dioxide to freeze food. conventional freezing is more challenging with larger foods such as potatoes and apples as compared to the small-sized foodstuffs, like kernels of sweetcorn and peas etc.

FOOD QUALITY

Drying of food results in food quality changes (Ratti C, 2001). The quality of foods can be divided into three sections: physical, chemical, and nutritional. Major qualities of foods that can be affected during drying processes are color, odor, texture, rehydration property, bulk properties, flow property, water activity, and retention of nutrients and volatile compounds (Sablani S.S, 2006). Regarding to nutritional qualities, oxygen, high temperature, and cell damage are usual enemies of bioactive compound retention during processing.

Goal of Freezing

1. To prevent growth of microbes by-
    a. Killing some bacteria
    b. Reducing water action
    c. Mechanical formation of ice crystals

   d. Osmotic changes in cell fluids
   e. Decrease the amount of free water
2. To lower temperature enough to slow down chemical reactions- (every 10 °C decrease in temperature that lead to half the reaction rate).

Types of Freezing

1. Fast Freezing- Quick freezing occurs at -25°C or less. Ice crystals are small and do not damage food cells.

Main features-

   a. Flash freeze
   b. 5-10-minute process
   c. Higher amounts of cryo-protectant
   d. Less chance of ice crystal formation
   e. Eggs warmed for fertilization
2. Slow Freezing- It occurs at-24° C or above. Ice crystals are big and damage the food cells causing loss of texture, nutrients, color and flavor on thawing.

Main features-

a. Gradual freeze
b. 3 hr. process
c. Lower amounts of cryo-protectant
d. Greater chance of ice crystal formation
e. Eggs thawed for fertilization

Advantages of quick freezing over slow freezing

1. Smaller ice crystals are formed, hence there is less mechanical destruction of intact cells of the food.
2. There is a shorter period of solidification and therefore, less time for diffusion of soluble materials and separation of ice.
3. There is more prompt prevention of microbial growth and there is more rapid slowing of enzyme activity.

Demerits of Quick Freezing

1. Higher investment in deep freezer or freezing equipment or frozen stores.
2. Thawing problems in some cases.
3. Need for refrigerated display cabinates for retial marketing.

4. Freezing technology at present suitable for upper and middle class.
5. Products are more prone to dehydration due to large exposed surface.
6. Small fluctuation in storage temperature can result in thawing into a large lump.

Supercooling

Going below freezing point without the formation of ice crystals (crystallization). It yields better quality food than if not present. It shows that the undesirable effects of freezing are due to ice formation rather than reduction of temperature.

The freezing point varies in function of the composition of the food, but is almost never lower than -5°C.

Examples-

Table- 1 Examples of Food and solutions commonly used in Supercooling

Product	°C
Milk, eggs	-0.5
Meat	-1.7 to -2.2
Fish	-0.6 to -2.0
Vegetables	-0.8 to -2.8
Fruit	-0.9 to -2.7
1 M sachralose solution	-2.65
1 M NaCl solution	-3.45

Effects of Supercooling in Food Industry

1. Supercooling phenomenon is of particular interest in food industry.
2. Fruits and vegetables storage and transport require cold environment where the temperature is several degrees below freezing point.
3. However, freezing the food lead reduce the quality by damaging food cells and changing color and sensory.

Ultrasound technology to assist freezing

This technique first produced internationally by Prof. Sun and his team and their experimental analysis showed that ultrasound freezing assists to preserve the internal structure of many vegetables, like potatoes, apples and numerous larger fruits etc. This technique has the capability to make cavitation, during the freezing process it can assists to decrease the size of ice crystals that form in the internal structure of the food. A recent report of Market Research Future (MRF) forecasted by 2027, there will be 751 million tons reach at 5.34% annual growth rate of global vegetable market and frozen fruits (L. Y. Zheng and Da-Wen Sun, 2006).

Vacuum Freeze- drying

Vacuum freeze-drying of biological materials is one of the best methods of water removal, with final products of highest quality. The solid state of water during freeze-drying protects the primary structure and the shape of the products with minimal volume reduction. The lower temperatures in the process allow maximal nutrient and bioactive compound retention. This technique has been successfully applied to diverse biological materials, such as meats, coffee, juices, dairy products, cells, and bacteria, and is standard practice for penicillin, hormones, blood plasma, vitamin preparations, etc. Despite its many advantages, having four to ten times more energy requirements than regular hot air drying, freeze-drying has always been recognized as the most expensive process for manufacturing a dehydrated product.

FREEZE DRYING (FD) OR LYOPHILIZATION

Freeze-drying method by lack of liquid water, oxygen-free environment (if operated under vacuum), and low operating temperatures is thus the best choice to dehydrate fruits and vegetables in order to keep an optimized bio compound content in the final products. Despite the long dryingtime and expensive process, freeze-drying is widely used to produce high-value food products due tomaximal retention of food quality when compared to other drying techniques. FD is considered as the standard or reference drying method in most research studies. It is known as lyophilization, is a well-known technique for the productionof high quality food powders and solids (Karam *et al.*, 2016). Use of FD to numerous fruits and some fruits like apple, guava, strawberry, blackberry and tomatoes (Shishehgarha*et al.*, 2002; Hawlader *et al.*, 2006; Franceschinis *et al.*, 2014).

Quick-freezing

It is a process, which utilizes a medium at very temperature (-90 or -80 C), which guarantees avoiding damage to interior food cellmembrane by swiftly converting water into tiny ice crystals (Kim *et al.*, 2015).Normally, very quick-freezing mechanism are obtained by the use of impingement technology (cryogenic fluid), According to work done by Agnelli and Mascheroni (Agnelli ME, Mascheroni RH, 2001), a cryogenic immersion freezer by utilizing liquid nitrogen (-196 C) fused with an air-blast freezer by utilizing air as a media was studied. The cryogenic immersion technology assists to raise the heat transfer coefficient, thus decreasing the freezing time, but its adoption is limited by the high cost of the liquid fluids and by the kind of food. Alternatively, an auspicious saving of the freezing time can be attained by impingement technology (Salvadori VO and Mascheroni RH, 2002). Freezing process at very low temperature -90/-80 C plays very significant part, allowing to obtain fast freezing processes for a Broader set of food products (Pham QT,1984).

Cryogenic freezing

Technical feature uses a fluid (LN2) with very low evaporation temperature (-196 C),very short freezing times, High final quality, Reduced weight losses, Lower investment costs, generally lower energy usage and High running costs (refrigerant is lost).

Immersion Freezing (IF)

Features-

1. Freezing of bare or packaged foods in anagitated aqueous solution.
2. Solution can be monophasic (liquid) orbiphasic (with suspended ice crystals).
3. Solutions may be binary (only one solute: salt,sugar or alcohol) or ternary (two types ofsolutes).

Merits-

1. Rapid heat transfer
2. Lower investment and operating costs
3. Better product quality
4. Possibility of product formulation
5. Environmental friendliness

Demerits-

1. Leaching of food components to the solution
2. Solute uptake from the solution
3. Undesired taste or flavour in frozen unpackaged Products
4. Reuse, recovery or disposal of solutions

De-hydro-freezing

The concept of de-hydro-freezing is not new, with published studies carried out since the 1950s. The mechanics behind this process is that since fresh fruits and vegetables contain more water than meat and their cellular structure is less elastic, reducing the water content prior to freezing reduces the freezing time, the initial freezing point and the amount of ice formed within the product. This potentially reduces damage and improves product quality. Reducing the water load. Combined preservation method with two stages:

1. Partial dehydration by Hot air drying and Osmotic dehydration.
2. Rapid freezing by Fluidized bed freezer and Belt freezer.

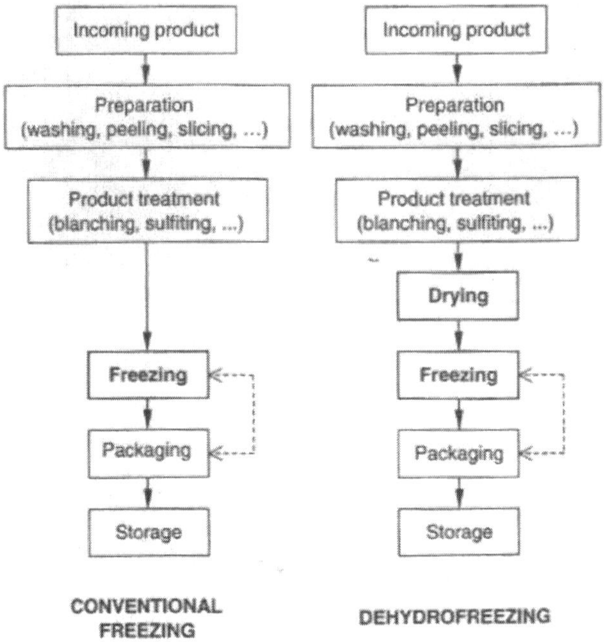

Fig. 1- Conventional versus De-hydro-freezing

OSMOTIC DEHYDRATION (OD)

OD is usually preferred over air-drying; such a process may be particularly suited to products destined for use as ingredients in processed foods like Soaking of a bare food product in a hypertonicaqueous solution, Loss of water and soluble components, Uptake of solutes from the dehydrating solution and in Solutes: Sugars, salts, alcohols, pure or in combination.

Merits-

1. Better product quality
2. Possibility of product formulation
3. Individual freezing
4. Lower investment and operating costs
5. Lower energy needs for freezing, storage and transport

Demerits-

1. Solute uptake
2. Undesired taste and flavour

3. Natural components leaching
4. Reuse and/or disposal of the dehydrating solution

Pressure-Shift Freezing (PSF)

Features-

1. Continuous lowering of freezing point at increasing pressures
2. Instantaneous nucleation and ice crystal growth during depressurization at low temperature
3. Homogeneous nucleation in all the volume (intra and extra cellular nucleation)
4. Numerous ice crystals with smaller average size

Demerits-

1. Very high investment costs
2. Discontinuous process
3. High operating costs
4. Possible protein denaturation in meats due to pressure

Radio-frequency assisted freezing (R-FAF)

Work has recently been published on the potential of radio frequency (RF) to assist food freezing. The proposed principle behind microwave and radio-frequency assisted freezing is similar to that of microwave assisted freezing, i.e. to exploit water dipole rotation induced by microwaves to disrupt ice nucleation and formation during freezing (Aneseet *al.*, 2012).

Microwave assisted freezing (MAF)

Hanyu*et al* and Jackson *et al.*, studied the potential of MAF to help in freezing, they propounded principle behind MAF is to fully utilization of water dipole rotation induced by microwaves to disrupt ice nucleation and formation during freezing. (Hanyu*et al.*, 1992; Jackson *et al.*, 1997).

Electrostatic assisted freezing (EAF)

The principle is that applying an electric field (E.F) to a food will orientate the polar molecules, such as water, thus controlling super-cooling and ice crystallization. It has not, as yet, been assessed in a food system (Woo *et al.*, 2010).

Magnetic resonance assisted freezing (M-RAF)

A Japanese patented system called CAS (Cells Alive System) freezing, which involves variable magnet fields, has been marketed by ABI (Japan) for enhanced freezing and frozen storage of foods. This technique claims to "retain the texture and flavour of food". by enhancing super cooling of the product, achieved by subjecting the target product with a low intensity magnetic field (Dodd J, 2004).

Ultrasonic assisted freezing (UAF)

Ultrasonically assisted thawing has been a proven process for a relatively long time, its use to assist freezing is more recent. Research has mainly utilized 'power ultrasound' a form of ultrasound waves with low frequency (18-20 kHz to 100 kHz) and high intensity (generally higher than 1 W/cm2) (Zheng L and Sun D W, 2005).

Features-

1. Ultrasound enhances vibration of molecules facilitating ice nucleation
2. Power Ultrasound is used (low frequency, high intensity)
3. Intra and extracellular crystals are produced simultaneously
4. Smaller ice crystals
5. Less damage to structure
6. Lower freezing time

USE OF PROTIENS IN FFT (FOOD FREEZING TECHNOLOGY)

1. Antifreeze proteins

The use of antifreeze proteins (naturally present in fish inhabiting in polar and northern coastal waters also been found in numerous invertebrates, plants and fungi). One application of antifreeze proteins that is in the food industry is in low-fat ice creams and frozen yogurts. The main function of antifreeze to suppress the growth of ice nuclei, thus inhibiting ice formation, manipulation of ice crystal growth and retarding recrystallization on frozen storage (Wang S and Sun D.W, 2012).

2. Ice-nucleation proteins

Such proteins are made by some Gram negative bacteria. The chief principle of ice-nucleation proteins is to increase the temperatures of ice-nucleation anddecrease the degree of supercooling. thus shortening the freezing time and aiding the nucleation of ice throughout the product (Li J and Lee T.C,1995).

Main Provocations in Food freezing technologies

- To find suitable apparatus and environments for freezing novel products.
- To develop novel apparatus to fulfil the actual and future needs in food processing: Attain the highest final quality, optimize energy usage, Reduce the environmental impact.

REFERENCES

- Biglia A, Fabrizio E, Ferrara M, Gay P, RicaudaAimonino D. Performance assessment of a multi-energy system in a food industry. *Energy Procedia* 2015; **82:**540-45.
- Dempsey P, Bansal P. The art of air blast freezing: Design and efficiency considerations. *ApplThermEng*2012; **47:**71-83.
- L. Y. Zheng and Da-Wen Sun, Innovative Applications of Power Ultrasound during Food Freezing Processes - A Review, Trends in Food Science & Technology, 17 [1] (2006) 16-23.
- Ratti, C. Hot air and freeze-drying of high-value foods: A review. *J. Food Eng*. 2001, **49:**311–319.
- Sablani, S.S. Drying of Fruits and Vegetables: Retention of Nutritional/ Functional Quality. Dry. Technol. 2006, **24:**123–135.
- Karam, M.C.; Petit, J.; Zimmer, D.; Djantou, E.B.; Scher, J. E_ects of drying and grinding in production of fruit and vegetable powders: A review. *J. Food Eng*. 2016, **188:**32–49.
- Shishehgarha, F.; Makhlouf, J.; Ratti, C. Freeze-Drying Characteristics of Strawberries. Dry. Technol. 2002,**20:** 131–145.
- Hawlader, M.N.A.; Perera, C.O.; Tian, M.; Yeo, K.L. Drying of Guava and Papaya: Impact of Different Drying Methods. Dry. Technol. 2006, **24**:77–87.
- Franceschinis, L.; Salvatori, D.M.; Sosa, N.; Schebor, C. Physical and Functional Properties of Blackberry
- Freeze- and Spray-Dried Powders. Dry. Technol. 2014, **32:**197–207.
- Kim Y, Hong GP. 2016. Effect of artificial supercooling followed by slow freezing on the microstructure and qualities of porkloin. Korean *J Food Scian***36:**650-655.
- Kim BYH, Liesse C, Kemp R, Balan P. Evaluation of combined effects of ageing period and freezing rate on quality attributes of loins. *Meat Sci*2015; **110:**40-50.
- Agnelli ME, Mascheroni RH. Cryomechanical freezing. A model for the heat trasnfer process. *J Food Eng*2001; **47:**263-70.
- Salvadori VO, Mascheroni RH. Analysis of impingement freezers performance. *J Food Eng*2002; **54:**133-40.
- Pham QT. An extension to Plank's equation for predicting freezing times for foodstuffs of simple shape. *Int J Refrig*1984; **7:**377-83.
- Anese, M., Manzocco, L., Panozzo, A., Beraldo, P., Foschia, M. &Nicoli, M. C. (2012) Effect of radiofrequency assisted freezing on meat microstructure and quality. Food Research International. 46:1, 50-54

- https://www.newfoodmagazine.com/article/8682/innovative-freezing-echnologies-for-foods/#:~:text=With%20pressure%20shift%20freezing%2C%20the,and%20homogenously%20throughout%20the%20food.
- Hanyu, Y., Ichikawa, M. & Matsumoto, G. (1992) An improved cryofixation method – cryoquenching of small tissue blocks during microwave irradiation. Journal of Microscopy-Oxford. **165:**255-271
- Jackson, T. H., Ungan, A., Critser, J. K. & Gao, D. Y. (1997) Novel microwave technology for cryopreservation of biomaterials by suppression of apparent ice formation. Cryobiology. 34:4, 363-372
- Woo, M. W. & Mujumdar, A. S. (2010) Effects of electric and magnetic field on freezing and possible relevance in freeze drying. Drying Technology. **28:**4, 433-443.
- Dodd, J. (2004) Out of the ice age: Technology to freeze by. Japan Inc. Communications. http://findarticles.com/p/articles/mi_m0NTN/is_57/ai _n6118032 (Accessed 17/12/08)
- Zheng, L. & Sun, D.W. (2005) Ultrasonic assistance of food freezing. Chpt 23, pp603-626. In: Emerging Technologies for Food Processing, edited by Sun, D.W. Elsevier Academic Press, London, UK
- Fikiin, K. (2003) Novelties of Food Freezing Research in Europe and Beyond. Flair-Flow 4 Synthesis Report. SMEs No. 10. Project No: QLK1-CT-2000-00040
- Wang, S. & Sun, D.W. (2012) Antifreeze proteins. Chpt 31, pp693-708. In: Handbook of Frozen Food Processing and Packaging, 2nd Ed, edited by Sun, D.W. CRC Press, Taylor & Francis Group, Boca Raton, Fl
- Li, J. & Lee, T.C. (1995) Bacterial ice nucleation and its potential application in the food industry. Trends in Food Science & Technology. 6, 259-265

9

CHAPTER

BIOTECHNOLOGICAL ROLE IN FOOD PROCESSING

Madhvi Sharma[*1], Fozia Hameed[2], Nadira Anjum[2] and Anjum Ayoub[2]

*[*1]Assistant Professor, PG Department of Biotechnology, Khalsa College, Amritsar*

[2]Division of Food Science and Technology, Sher-e-Kashmir University of Agricultural sciences and Technoloy-Jammu, Chatha-180009, India

*Corresponding Author: [*1]madhvisharma413@gmail.com*

ABSTRACT

Biotechnology as a discipline deals with the applicability of multiple living beings to the production of usable goods. Recently, time has seen the implementation in the area of flavour improvement as food additives and the production of a broad variety of value-added products. This is the key purpose of the use of biotechnology is to produce genetically modified food and to grow genetically modified starts. Biotechnology: Significant developments in the world of food manufacturing have taken place in the production of goods with longer shelf life.

Keywords: Biotechnology; Enzymes; Genetic Modification, Food Processing; bioengineering.

INTRODUCTION

Biotechnology may be characterized as the use of living organisms to produce usable goods. It shall also be defined as any technical application that uses biological beings, biological organisms or their derivatives for the manufacturing or transformation of products for some uses. It overlaps time again with the biomedical engineering and bioengineering areas, based on the various technologies and methods used. Applications of biotechnology in food processing requires the following applications, such as – Usage of fermentation of substances in order to improve the properties of the material products such as flavour, fragrance, shelf life, texture, quality and nutritional value of the specific food product.

Biotechnology plays an important role in the development of enzymes and the use of such enzymes contributes to the requisite improvements in food. Biotechnology is used in the processing of foodstuff constituents; flavourings, sweeteners, food additives and a number of other ingredients; highly regarded goods, genetically modified organisms and with the seeds.

Biotechnology in the food industry has been the core subject of several research studies, national and international symposia, and other significant reference works in recent years (Earle, 1984; Harlander and Labuza, 1986; Jarvis and Holmes, 1982; Kirsop, 1985; Knorr, 1987; Knorr and Sinskey, 1985; Moo-Young et al., 1985; Rehm and Reed, 1983). Reports of significant advances have come from the full spectrum of biotechnology research and development resources: universities and institutes as well as genetic "biotiques" and large food corporations. Important business alliances continue to be formed on a worldwide scale, linking advanced biotechnology research skills with large producers and marketers of food products. Biotechnology aims to boost food production, increases food production, harvesting index, nutritional and storage value is also enhanced, enhanced production of raw materials, enriched flavourings, and the production of food containing vaccines. Biotechnology could have a dramatic impact on the entire agriculture and food sector. It has the potential to reduce the need for agricultural chemicals; improve the productivity, efficiency, and profitability of food production and processing; open new markets for improved or unique processed food products; and, improve the nutritional quality, safety, cost, and convenience of consumer food products. Any improvement of the food supply at any point in the food chain will ultimately impact the utilization end of the system and the ultimate beneficiary of the improvement — the consumer.

Biotechnology as a source of enzymes production

A latest U.S. Enzyme Demand Survey estimated gross sales of $185 million in 1985, 58% of which were in the food industry (Charles-Kline and Co., 1986). Of the types of enzymes used by the food industry, proteases and carbohydraces account for much of this market. The prevalent protease marketed is rennin (chymosin), which is used in the cheese-making process to coagulate milk to form curds. Carbohydraces, those used in the production of cornstarch (so-called amylase, glucoamylase and glucose isomerase starch enzymes) account for 85 per cent of sales. Other enzymes have a number of uses, including flavouring production e.g., lipase in cheese making) (Arbige et al., 1986), enhanced extraction (e.g., pectinase in juice processing) (Kilara, 1982), and improvements in food functionality (e.g., -amylase in delayed bread stalling) (Boyce, 1986). The use of enzymes is carried out at the industrial stage of food production items as well as improving their efficiency. Food manufacturing industry makes use of the enzymes that are generated worldwide with the aid of genetically engineered species. The enzyme, thus made consists of carbohydrates and proteases. These enzymes had number of uses, such as cheese processing, curd making and the addition of flavours to food products. In developing countries, the largest fraction of the enzymes used in the food industry are proteins and carbohydrates, which make up more than 50% of the enzymes used (Lawrence et al. 1988). Industrial processing of enzymes requires, in particular, utilisation of the micro-organisms. Microorganisms are cultivated in large containers after which the desired enzymes are secreted into the medium in which the microorganism has been fermented. Enzymes are secreted as a result of microbial interaction in the form of the metabolites. Enzymes developed with such mentioned process is then removed, the purification process is underway and these distilled enzymes are further used in food processing, nutrition business and a number of other applications. Enzymes purified do not consist of a cell or some other type of macromolecules such as DNA or the RNA.

The microorganism was licenced for use by the British Ministry of Agriculture, Fisheries and Food on 1 March 1990. The maker, Gistbrocades, was given permission to produce and sell the products. The strain of yeast, *Saccharomyces cerevisiae* 352 Ng, used in the baking industry. The strain was genetically modified to develop elevated levels of two enzymes involved in the use of starch, maltose permease and maltase. Stable microbes about 70% of the world's population were losing the failure to ferment lactose due to the slow depletion of lactase and the enzyme present in the intestinal boundary of the brush that hydrolyzes lactose, glucosa and galactose. Several investigators have found that lactase-deficient individuals absorb yoghurt lactose even more effectively (Savaiavo and Levitt, 1987).

Probiotics: Various bacterial strains are capable of being inserted and they compete in the gastrointestinal system of humans and livestock. These organisms are also known as "probiotic" because their role is to support host in a way that is beneficial. Construction of strains capable of competitively inhibiting potentially pathogenic intestinal species may have many uses in agriculture and food. For example, nonpathogenic strains of *Salmonella* or other gut organisms could be engineered to produce broad spectrum bacteriocins. These strains could be supplied in poultry feed and water for biological control of pathogenic strains of *Salmonella* or other gut pathogens. *Lactobacillus acidophilus*, a food-grade microorganism used in the processing of acidophilus milk, is capable of surviving passage through the stomach and under some conditions, is capable of colonising the GI tract. It may be necessary to engineer bacteriocin-producing strains with improved colonisation capacity to modulate intestinal ecology.

Microbially produced ingredients

Microorganisms contain a variety of substances metabolites currently used as additives in processed foods that include acidulants (acetic, lactic, benzoic, propionic), flavourings (diacetyl, pyrazines, lactones, esters), flavour enhancers. (MSG), pigments (monascin, astazanthane), stabilisers and thickeners (xanthan gum, dextrans), nutritive additives (vitamins, amino acids), sweeteners (aspartame), enzymes (proteases, lipases, cellulases, pectinases) and preservatives (nisin).These ingredients add functionality, enhance nutritional quality, extend shelf-life, improve convenience and ensure safety (Neidleman, 1990).

Many of the genes for these enzymes have been cloned so as to get higher production in less time period. These enzymes are used for making cheese, curd and flavoring food items. Major percentage of these enzymes is used in food industry as in US more than 50% of proteases and carbohydrases are used in food industry. These enzymes include rennin and -amylase.

α-Amylase: High Fructose Corn Syrup Industry

The first petition to the Food and Drug Administration (FDA) to validate the GRAS status of a food-processing enzyme developed by recombinant DNA techniques was for -amylase. This pioneering petition was filed by CPC International, Inc. on 9 July 1984. -Amylase is the enzyme used in the first step in the production of high-fructose corn syrup (HFCS), a widely used nutritive sweetener derived from cornstarch.

Chemosin (Rennin): the dairy industry.

Another enzyme that has been the subject of significant genetic engineering research is chymosin, an active rennet ingredient used in the dairy sector to coagulate milk to form curds in the cheese-making process. Chymosin is an endoprotease that is particularly specific to the hydrolysis of peptide in bonds in the V-fold of kappa-casein milk, resulting in the destabilisation of casein micelles and eventual curd development. The propensity for microbial development of calf rennet chymosin has led a number of companies to develop strategies to clone the gene for chymosin from cDNA libraries extracted from calf stomach mRNA and to achieve heterologous gene expression in different host species (McGuire, 1986).

Some genetically modified enzymes used in food industry:

- Catalase used in mayonnaise production and it removes hydrogen peroxide.
- Chymosin useful in cheese production as it coagulates milk.
- Glucose oxidase is used in baking as it stabilizes the dough.
- α-amylase converts starch into maltose and used in baking for sweetness.
- Protease used for meat tenderization process, baking and dairy products (Snow, 1985).

Fermentation

Enzymes are used in the cultivation and distribution of food products specifically manufactured on an industrial basis. From the past decade till the present century food manufacturing firms are using enzymes. They are created by genetically modified organisms (European Food Knowledge Committee for 2015). These enzymes are made up of proteins and carbohydrase. Breweries are synthesised by the fermentation process. 'It is used to make breweries at an industrial level for the manufacture of bilious yeast strains. Genetic engineering has helped us to make light wine. It's Yeast Genetically engineered by international gene coding for glucoamylase. During the fermentation process, the yeast expresses glucoamylase, convert starch to glucose (Lawrence et al. 1988)

Wine Making

Snow (1985) suggested a technique for the genetic modification of industrial yeast strains used in wine making to develop the potential for malolactic fermentation. The primary fermentation that occurs in wine production is accomplished by the use of yeast to turn sugar to alcohol. Secondary fermentation can occur, in particular during the production of red wines catalysed by bacteria in the genera *Leuconostoc, Lactobacillus* and *Pediococcus*. During this secondary fermentation, malic acid is converted to lactic acid, which decreases the acidity of the wine, increases the acidity of the final wine and making the taste more complex. Procedures used to promote malolactic fermentation can increase the risk of loss of quality of wine. They also raise the cost of wine production.

Biotechnology: Improving Food Nutrition and Enhancing Taste

Not every food item includes all the necessary components. That's why not every food item has optimal nutrition. For instance in many countries of the world, rice is used as a staple meal. Not to be devoid of Vitamin A is not the perfect staple food. Usage of biotechnological good techniques has overcome these issues through the implementation of international technologies Gene of Vitamin A (Sun SS 2008).

Biotechnology has paved the way for scientists to grow quality fruit taste. GM foods with a better taste include seedless melon, tomato, eggplant, pepper, cherries, etc. Elimination of seed from food articles have resulted in a more soluble sugar content that raises sweetness (Falk, 2002). By the year 2050, population of world will become nine billion. So more yield will be requiring on same land. Biotechnology is potentially best technology to fight against problem of food yield (http://cals.arizona.edu/pubs/health/az1066).

Biosensors

Extremely specific activities of enzymes and microbial cells may be manipulated as analytical methods for calculating the concentration of particular components in complex mixtures. Enzymes, antibodies, or entire cells can be immobilized. It can be observed on solid surfaces and the particular reactions they induce electrochemically, photometrically, thermometrically or mechanically (Wagner and Schmid, 1990). In food systems, biosensors can be used to quantify low molecular weight, single compounds such as glucose, organic acids, amino acids, alcohols or food additives; complex compounds such as microorganisms and biological or chemical contaminants; and complex consistency parameters such as freshness, prevalence of shelf-life, taste, maturity.

Table: Commercially available biosensors for food analysis:

S.No.	Analyte	Biocomponent	Application
1.	Glucose	Whole bacterial cells Glucose oxidase enzyme	Molasses production Brewing, various fermentations, fruit juice and soft drink manufacture, banana maturation
2.	Lactose	P-galactosidase enzyme	Raw milk
3.	Sucrose	Invertase enzyme	Instant cocoa manufacture
4.	Lactate	Lactate dehydrogenase enzyme	Dairy products, yogurt, whey
5.	Ethanol	Alcohol dehydrogenase enzyme	Alcoholic beverages, wine, beer, cider, fermentations
6.	Peptides	Amino peptidase enzyme	Casein hydrolysis
7.	Amino acids	Aminoacid dehydroge-nase enzyme	Many foods
8.	Gluta-mate	L-glutamate oxidase enzyme	Soy sauce manufacture
9.	Aspar-tame	L-aspartase enzyme or alcohol oxidase en-zyme	Level of sweetener in many foods including soft drinks
10.	Ascorbic acid	Ascorbate oxidase enzyme	Fruit juices
11.	Sulfite	Sulfite oxidase enzyme	Dry fruit, wine, vinegar, juices, potato flakes
12.	Penicillin	Antibody-enzyme conjugate	Milk
13.	PHB ester	phydroxybenxoate hydroxylase enzyme	Fruit juices and drinks

(Wagner and Schmid, 1990).

CONCLUSIONS

The food industry is characteristically pragmatic in the amount it spends in innovation (usually 1.5% or less of sales) and in the implementation of increasingly developing fields of technology. Biotechnology research, as it currently operates, is largely funded by traditional power bases such as DuPont and Monsanto, as well as by a growing number of small research firms, biotics. Many of these small businesses are highly qualified in their fields of specialization and are highly motivated, sensitive and resourceful. What we see in food biotechnology is the successful use of research partnerships between major food processors and biotics. This has enabled food corporations to rapidly achieve critical mass in specific areas of research and has helped to drive advances in this field. Several of the examples used in this paper were the product of these treaties. This phenomenon is continuing to develop. As food companies become more familiar with the technology and begin to experience its success in the marketplace, we will see internalization of research skills and the full integration of biotechnology into the well-established food research disciplines.

The consequence is the growth of major companies that are horizontally organized across a vast variety of food industries. As this happens, along with the internationalization of biotechnological science expertise, we will witness a strong change towards vertical integration in the direction of our raw material base, which will place the "value-added cascade" at the genetic level to start further back in the system. This may mark the advent of a significant new competitive fulcrum.

REFRENCES

- Arbige, M.V., P.R. Freund, S.C. Silver, and J.T. Zelko. 1986. Novel lipase for cheddar cheese flavor development. Food Technol. 40:91-96, 98.
- Boyce, C.O.L. 1986. Novo's Handbook of Practical Biotechnology. Novo Industri A/S, Bagsvaerd, Denmark. 125 pp.
- Charles Kline & Co. 1986. American market for enzymes expected to reach $260 million by 1990. Genet. Eng. News. 6:18.
- Earle, R.L. 1984. Food engineering and biotechnology. Food Technol. Aust. 36:498.
- Falk MC, Chassy BM, Harlander SK, Hoban IV TJ, McGloughlin MN, et al. (2002) Food biotechnology: benefits and concerns. J Nutr 132: 1384-1390.
- Harlander, S.K., editor; , and T.P. Labuza, editor. , eds. 1986. Biotechnology in Food Processing. Noyes, Park Ridge, N.J. 349 pp.
- Jarvis, B., and A.W. Holmes. 1982. Biotechnology in relation to the food industry. J. Chem. Technol. Biotechnol. 32:224-232.
- Kilara, A. 1982. Enzymes and their uses in the processed apple industry: A review. Process Biochem. 17:35-41.
- Kirsop, B. 1985. Opportunities for biotechnology in food processing. Pp. 175-192 in G.E. Russell, editor. , ed. Biotechnology and Genetic Engineering Reviews, Vol. 3. Intercept, Dorset, England. Distributed by Scholium International, Great Neck, N.Y.
- Knorr, D., and A.J. Sinskey. 1985. Biotechnology in food production and processing. Science 229:1224-1229. [PubMed]
- Knorr, D.W., editor; , ed. 1987. Food Biotechnology. Marcel Dekker, New York. 625 pp.

- Lawrence R. New applications of biotechnology in the food industry. Biotechnology and the Food Supply: Proceedings of a Symposium (1988).
- McGuire, J. 1986. Cloning, expression, purification and testing of chymosin. Pp. 121-141 in BIO EXPO 86: Proceedings of the American Commercial & Industrial Conference & Exposition in Biotechnology. Organized and presented by Cahners Exposition Group, April 29-May 1, 1986, Boston, Mass. Butterworth, Stoneham, Mass.
- Moo-Young, M., editor; , H.W. Blanch, editor; , S. Drew, editor; , and D.I.C. Wang, editor. , eds. 1985. Comprehensive Biotechnology: The Practice of Biotechnology: Current Commodity Products, Vol. 3. Pergamon, Elmsford, N.Y. 1136 pp.
- Neidleman, S. (1990) The microbial production of biochemicals. In: A Revolution in Biotechnology. Marx, J.L. (Ed.). Cambridge University Press, New York, N.Y.
- Rehm, H.J., editor; , and G. Reed, editor. , eds. 1983. Biotechnology, A Comprehensive Treatise in Eight Volumes: Food and Feed Production with Microorganisms, Vol. 5. VCH, Deerfield Beach, Fla. 473 pp.
- Scowcroft, W.R., R.I.S. Brettell, P.A. Davies, P.J. Larkin, M.A. Pallotta, S.A. Ryan, and R. Appels. 1985. Somaclonal variation -- biology and agricultural impact. P. 4 in G.A. Galau, editor. , ed. Abstracts, First International Congress of Plant Molecular Biology. Organized by the International Society for Plant Molecular Biology, Oct. 27-Nov. 2, 1985, Savannah, Ga. Center for Continuing Education, University of Georgia, Athens, Ga.
- Snow R (1985). Genetic engineering of a yeast strain for malolactic fermentation of wine. Food Technol 39: 98-101.
- Sun SS (2008) Applications of agricultural biotechnology to improve food nutrition and health care products. Asia Pac J Clin Nutr 17: 87-90.
- (http://cals.arizona.edu/pubs/health/az1066).
- Wagner, S. and Schmid, R.D. (1990) Biosensors for food analysis, Food Biotechnology 4:215.

10

CHAPTER

HURDLE TECHNOLOGY

Anjum Ayoub[*1], Kundan Veer Singh[2], Jagmohan Singh[1], Fozia Hameed[1] &Nadira Anjum[1]

[1]*Division of Food Science and Technology, Sher-e-Kashmir University of Agricultural sciences and Technoloy-Jammu, Chatha-180009, India*

[1]*Division of Plant Breeding and Genetics, Sher-e-Kashmir University of Agricultural sciences and Technoloy-Jammu, Chatha-180009, India*

*Corresponding Author: [*1]anjumparay77@gmail.com*

ABSTRACT

People's demand for food with minimal changes in sensory and nutritional attributes has led to the emergence of hurdle technology in the field of food technology in the modern era of growth. Hurdle technology is a novel approach to the preservation of both plant and animal-based foods. Many countries around the world, including India, use this technology. Food safety factors are referred to as obstacles. Heat, refrigeration, preservatives, irradiation, and other barriers are used to prevent or remove the growth of microbes that may spoil food or trigger food poisoning. To control organisms, at least two or three hurdles must be included in the scheme. Each barrier creates a hostile environment for microorganisms, preventing them from multiplying in the food and thereby extending the shelf life of the products. It is critical that the right set of barriers be in place to ensure complete freedom from microbes and, as a result, ensure microbial protection, stability, organoleptic, and nutritive quality of the food. Mountaineers and space scientists will benefit from hurdle technology by receiving stable and ready-to-eat meat items. Food protection with several targets may be a cost-effective strategy in the food processing industry. More hurdles should be established that can be used safely and efficiently to improve the quality and protection of different foods. It is suggested that hurdle technology may be a critical component of food preservation in the future.

Key words: Hurdle technology, Food Preservation, Microbes, Safety, quality

INTRODUCTION

It has been projected that one third of the total harvested foodstuffs are lost and damaged before it is used. Hence, it is very significant to preserve food in order to avoid the huge losses occurring due to deteriorative changes brought by microbial, chemical, and physical process. Earlier preservation techniques are based on sole parameter that alters in nutritional quality and sensory quality of the food, but this hurdle technique brings minimum nutritional and sensory alterations in the foodstuff that makes the product more satisfactory and valuable than obtained by conventional ways. The chief norm of preservation is to create undesirable conditions for the growth of microbes in food (Pal, 2014). Conventionally, fermented sea foods, which are commonly used in Japan, are a good example of hurdle technology. Hurdle technique or Combined method technology (CMT) is the combined use of several preservation methods to make a product shelf-stable, to improve quality and to provide additional safety. Leistner in 1976, introduced the term "hurdle effect".According to Pal,2014, hurdles are the microbial safety and stability of most foodstuffs that are based on a use of preservation factors and Leistner, 2000 defined Hurdle technology as an intelligent combination of hurdles, which secure the microbial safety and stability as well as retains the nutritional benefit, organoleptic quality, and economic viability to food products. In recent years, food preservation and food security have become an important factor of concern. In modern era of food processing, the demand for fresh & good quality food products has led to the emergence of hurdle technology. Hurdle technology is an efficient technique for improving the quality of food and to enhance its shelf life. Hurdle technology is a method of ensuring that food products will be safe for consumption and extending their shelf life. It also ensures that pathogens in food products will be eliminated. Hurdle technology is applied in food industry for the gentle but effective preservation of foods. Hurdle technology combines non-thermal food processing with traditional or other emerging technologies. The most promising combinations include non-thermal methods, such as high hydrostatic pressure, ultrasound, and pulsed electric fields. In the inactivation of spores, it is necessary to use a combined methods approach using 'hurdles.' Hurdles are physical or chemical parameters that can be adjusted to ensure the microbial stability and safety of foods. The physical parameters include the processing and storage temperatures, water activity, pH, and redox potential at levels that inhibit or inactivate the microorganisms and thus render the food safe. Hurdle technology has been made to decrease the habit of preservatives in foods, and contains combined effect of hurdles to establish an additional antimicrobial effect, thus improving the quality of the food (Leistner, 1985).

AIMS OF FOOD PRESERVATION

Pal, 2014, presented numerous aims of food preservation

1. Warrant the safety of food from microbes
2. Avoid food spoilage
3. Increase the storage quality of food
4. To govern foodborne intoxications and infections
5. To expand the storage life of food
6. To decrease economic losses

Necessity for hurdle technique

The need for treated food has raised in recent times, with the fast growing economy. Moreover, consumers now a days are demanding fresh, natural and minimal processed ready to eat food products. Ongoing trend has been eating out and to consume ready to eat foods. New ecology routes for microbial growth have emerged.

Principles of Hurdle Technology

1. Microorganisms should not be able to "jump over" all the hurdles present in the food product.
2. Preservative factors prevents microorganisms from multiplying and causing them to remain inactive or even die.
3. The hurdle concept illustrates that complex interactions of temperature, water activity, pH etc. are significant to the microbial stability.

ESSENTIAL FACETS OF HURDLE TECHNIQUE

The requirement for quality foodstuffs has led to the rise of hurdle technique. Numerous steps are decisive to hurdle technique. There are four major mechanisms by which hurdle technique affects the growth of microorganisms in foods. The hurdle technology affects the physiology and growth of microorganisms in food. There are mainly 4 major mechanisms by which hurdle technology affects the growth of microorganisms in foods:

1. Homeostasis

Homeostasis is the constant ability of organisms to maintain a stable and balanced internal temperature. (Raso*et al.*,1998). The interference with homeostasis of microbes form an attractive and logical focus for improvement if food preservation techniques (Gould,1995). The idea behind homeostasis is already known in higher organisms (Leistner, 2000). Disturbing the homeostasis of the microorganisms by various hurdles eventually results in the death of the spoilage causing microbes thereby protecting the food product from microbial spoilage.

2. Metabolic Exhaustion

The microbes in hurdle treated stable products use their energy for homeostasis, thereby, become metabolically exhausted. This leads to auto-sterilization of food products. Hence, microbiologically stable food becomes safe for keeping at ambient room temperature (Pundir and Murtaza, 2015).

3. STRESS REACTION

It is a very significant feature for efficient and effective preservation of targeted food material.

Biosynthesis of protective stress shock proteins (SSPs) is encouraged by numerous reasons, like water activity, pH, heat, ethanol etc. Contact to various stresses can cause the organism metabolically weak. Therefore, multi-target preservation of foods can be the key to avoid synthesis of stress shock proteins (Leistner, 2000).

4. Multi-target Preservation

This idea was first introduced by Leistner (1995). The multi-target preservation of food in which intelligently applied gentle hurdles will have a synergestic effects. application of numerous hurdles one by one would lead to an optimal microbial stability and effective food preservation (Leistner,1994).

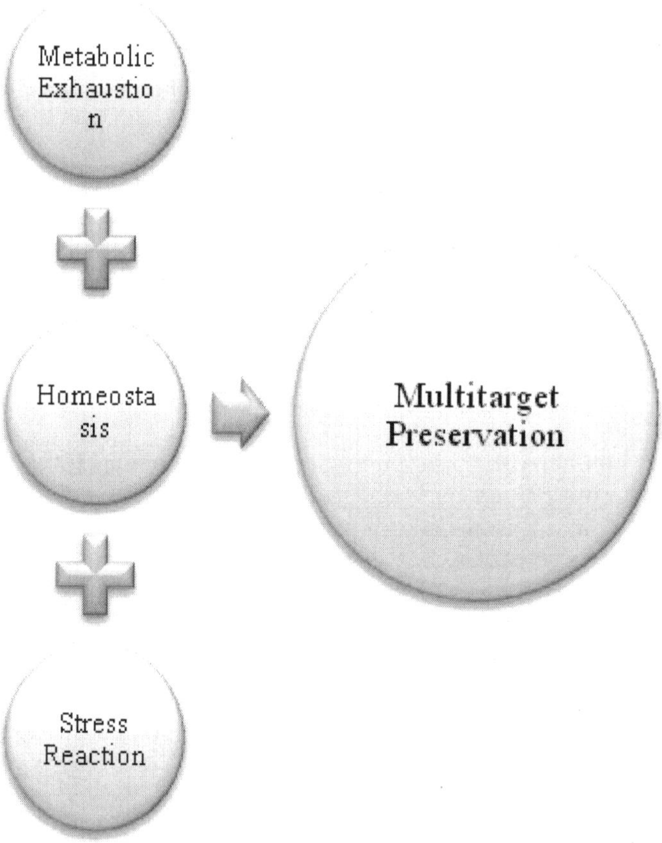

Fig. 1 Essential Facets of Hurdle Technique

Kinds of Hurdles

To govern the microbes in foodstuff, organic acids or common salt can be applied as hurdle.

Each hurdle aims to uproot, deactivate or hinderunwanted organisms from the foodstuff for high safety and food quality to the buyer. Numerous ordinary natural anti-microbials likenisin, natamycin, and other bacteriosins, and essential oils derived from rosemary also work well (Leistner, 1995; Pal *et al.*, 2014).

Table 1- Kinds of Hurdle utilized for Food Preservation

Kind of Hurdle	Examples
Physical	Aseptic packaging, electromagnetic energy (microwave, radio frequency, pulsed magnetic fields, high electric fields), high temperatures, (blanching, pasteurization, sterilization, evaporation, extrusion, baking, frying), ionizing radiation, low temperature (chilling, freezing), modified atmospheres, pack-aging films (including active packaging, edible coatings), photodynamic inactivation, ultra high pressures, ultra-sonication, ultraviolet radiation.
Microbial	Antibiotics, bacteriocins, competitive flora, protective cultures.
Physico-chemical	Carbon dioxide, ethanol, lactic acid, lactoperoxidase, low pH, low redox potential,low water activity, Maillard reaction products, organic acids, oxygen, ozone, phenols,phosphates, salt, smoking, sodium nitrite/nitrate, sodium or potassium sulphite, spicesand herbs, surface treatment agents.

(Source: Ohlsson and Bengtsson, 2002)

TABLE 2: Principal Hurdles used in Food Preservation

Parameter	Symbol	Uses
High temperature	F	Freezing
Low temperature	T	Freezing and Chilling
Reduced water activity	aw	Drying, curing, conserving
Increased acidity	pH	Acid addition or formation
Reduced redox potential	Eh	Removal of oxygen or addition of ascorbate
Bio-preservatives	-	Competitive flora such as microbial fermentation
Other preservatives	-	Sorbates, sulfites, nitrates

Source:*Leistner (1995)*

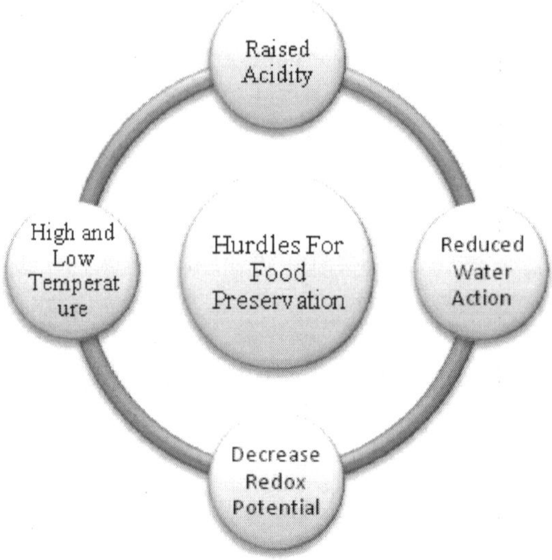

Fig. 2- Most Common Hurdles in Food Preservation

UTILIZATION OF HURDLE TECHNOLOGY IN DIFFERENT TYPES OF FOODS

In Food Industry Hurdle technology has been used in a broad kinds of foods viz., milk and dairy products, meat and meat products. poultry and poultry products, fish and sea foods, canned products, bakery products, juices, jams, pasta, salad, spices etc. in order to raise their shelf life (Leistner, 2000; Pal, 2014; Pundir and Murtaza, 2015). Application of several hurdles simultaneously would lead to an optimal microbial stability and effective food preservation (Leistner, 1995). The ginger has been identified as probable hurdle for enhanced storage life of curd rice(Famous South Indian traditional Indian dairy product) by 7 days at 37-degree C and 12 days at 4-6-degree C in comparison to earlier storage life of 24 hrs. at 30-degree C. (Balasubramanyam *et al.*, 2004). Application of 1% each of sodium chloride, sucrose and glycerol to reduce the water action of paneer, enhance storage life of Cottage Cheese curry (Rao and Patil, 1999). Numerous hurdles viz., antimicrobials, heat process and water action are used to preserve high moisture fruit products ex.- peach, pineapple, papaya, mango and banana (Alzamora*et al.*, 1993). Lee (2004) explained that hurdle technique can raise the microbial safety of vegetables and pickled fruits. According to Saxena *et al.*, 2009, infrared drying, osmotic dehydration and Gamma radiation can encourage the load of microbes in pineapple slices and thereby increasing storage life up to 40 days. According to Thomas et al., 2009, Establishment of hurdle technique is useful, as it enhanced the storage life of pork sausages. Sankhla *et al.*, 2012utilized hurdle technology in sugarcane juice by using numerous hurdles viz., food preservatives, heat treatment, irradiation and packaging materials. According to Gunathilake, 2006, Hurdle technique can be used in the preservation of fresh scrapped coconut using heat treatment and additives such as humectants and preservatives. Himalayan region yak milk product Dudhchurpi has self-life for several months without refrigeration. Most significant fordudhchurpi is its texture (elasticity), since people living at high altitude chew it as 'energy tablets'. Hossain applied numerous hurdles like heating, acid coagulation, addition of sugar and sorbate, smoking, drying and packaging in a closed container (Hossain, 1994). The experiments conducted by these researchers clearly indicated that the hurdle treated foods have better quality and long shelf life. The successful application of these hurdles shows that these can be recommended for preservation of all kinds of food products to enhance the product safety as well as stability.

BOON OF MULTI- HURDLE APPROACH OR COMBINED METHOD TECHNOLOGY

Boons of hurdle concepts are

1. It can avoid the severity of one hurdle for preservation
2. It can give synergy of combination
3. Many of the hurdles come from past experience (i.e. tradition or culture). Currently huge numbers of products are being developed based on the multi-hurdles
4. Improves product quality and microbial safety
5. Save Money, Energy & Several other Resources
6. Foods remains stable & safe, high in sensory & nutritive value due to gentle process applied
7. Doesn't effects the integrity of food pieces (fruits)
8. Applicable in both Large & Small Industries
9. Manufacture of new products according to the need of processors and consumers.

Conclusion

The requirement of the today's consumer to enjoy the safe and nutritious food that is much more fresh like can be fulfilled using hurdle technology concept. The hurdles collectively control microbial spoilage and food poisoning leaving desired fermentation processes unaffected. The novel goal for optimal food preservation is the multi-target preservation of foods applying various hurdles those exhibit a synergistic effect. Hence, hurdle technology is used in industrialized as well as in developing countries for the gentle and effective preservation of foods using two or more hurdles to an optimal extent without compromising the sensory quality of the food. Thus, hurdle technology will also be the key to future food preservation.

REFERENCES

- Leistner L (1985) Hurdle technology applied to meat products of the shelf stable product and intermediate moisture food types. In: Simatos D, Multon JL (eds) Properties of water in foods in relation to quality and stability. Martins Nijhoff Publishers, Dordrecht, pp 309–329
- Pal, M. 2014. Preservation of various foods. Ph.D. Lecture, Addis Ababa University, College of Veterinary Medicine and Agriculture, DebreZeit, Ethiopia. Pp.1-11. Pal, M., Gebretensey, A., Shiberu, T., Abdurahman, M. and Karanfil, O. 2014. The role of bacteriosin as food preservative. Beverage Food World **42:** 28-30.
- Leistner, L. 2000. Basic aspects of food preservation by hurdle technology. *International Journal of Food Microbiology***55:** 181-188.
- Raso, J., Barbosa-Canovas, G.V. and Swanson, B.G.1998. Sporulation temperature affects initiation of germination and inactivation by high hydrostatic pressure of Bacillus cereus. *Journal of Applied Microbiology* **85:** 17-24.
- Gould, G.W. 1995. Homeostatic mechanisms during food preservation by combined methods. In "Food Preservation by Moisture Control: Fundamentals and Applications". edited by G.V.Barbosa-Canovas and J.WeltChanes. Technomic, Lancaster, Philadelphia.
- Pundir, A. and Murtaza, N. 2015. Hurdle technology: An approach toward food preservation. *International Journal of Current Microbiology and Applied Science***4:** 802-809.
- Leistner, L., Gorris, L.G.M., 1995. Food Preservation by Hurdle Technology. *Trends Food Science and Technology* **6:** 41 46.
- Leistner, L., 1994. Further developments in the utilization of hurdle technology for food preservation. *Journal of FoodEnggineering* **22:**421 432.
- Leistner L and Gorris LGM. 1995. Food Preservation by hurdle technology. *Trend inFood Science Technology.,* **6:** 35-37
- Ohlsson T and Bengtsson N (2002) "The hurdle concept "Minimal Processing Technologies in the Food Industry, Woodhead Publishing, pp. 175–195. ISBN 978-1-85573-547-7.
- Leistner, L. 2000. Basic aspects of food preservation by hurdle technology. *International Journal of Food Microbiology* **55:** 181-188.
- Balasubramanyam, B.V., Kulkarni, S., Ghosh, B.C. and Rao, K.J. 2004. Application of hurdle technology for large scale production of curd rice. Annual Report. National Dairy Research Institute (Southern Campus), Banglore, India.

- Rao, K.J. and Patil, G.R. 1999. Development of ready to eat paneer curry by hurdle technology. *Journal of Food Science and Technology* **36:** 37-41.
- Alzamora, S.M., Tapia, M.S., Argaiz, A. and Welli, J. 1993. Application of combined methods technology in minimally processed fruits. *Food Research International* **26:** 125-130.
- Lee, S. 2004. Microbial safety of pickled fruits and vegetables and hurdle technology. *International Journalof Food Safety* **4:** 21-32.
- Saxena,S., Mishra,B.B., Chander, R. and Sharma,A.2009. Shelf stable intermediate moisture pineapple (*Ananascomosus*) slices using hurdle technology.*Journal of*Food *Science and Technology* **42:** 1681-1687.
- Thomas, R., Anjaneyulu, A.S.R., Kondaiah, N., 2010. Quality of Hurdle Treated Pork Sausages during Refrigerated (4± 1 C) Storage. *Journal of Food Science and Technology***47:**266-272.
- Sankhla, S., Chaturvedi, A., Kuna, A., Dhanlakshmi, K., 2012. Preservation of Sugarcane Juice Using Hurdle Technology. *Sugar Technology***14:**26-39.
- Gunathilake, K.D.P.P., 2006, Application of hurdle techniques to preserve fresh scraped coconut at ambient and refrigerated storage. *Journal of National. Science Foundation Sri Lanka***33:** 265-268.
- Hossain SKA. (1994). Technological Innovation in Manufacturing DudhChurpi. Ph.D. Thesis, University of North Bengal, Siliguri, India, pp.122.

11

CHAPTER

NANOTECHNOLOGY IN FOOD PROCESSING

Anjum Ayoub[1], Kundan Veer Singh[*2], Julie D.Bandral[1], Fozia Hameed[1] & Nadira Anjum[1]

[1]Division of Food Science and Technology, Sher-e-Kashmir University of Agricultural sciences and Technoloy-Jammu, Chatha-180009, India

[2]Division of Plant Breeding and Genetics, Sher-e-Kashmir University of Agricultural sciences and Technoloy-Jammu, Chatha-180009, India

*Corresponding Author: *2kundanveer@gmail.com*

ABSTRACT

Nanotechnology has revolutionised a variety of scientific and industrial fields, including the food industry. Food processing, food packaging, functional food production, food protection, detection of foodborne pathogens, and shelf-life extension of food and/or food products are just a few of the applications of nanotechnology that have emerged as a result of the growing need for nanoparticles in various fields of food science and food microbiology. This analysis summarizes the potential of nanoparticles in the food industry, with the goal of providing consumers with clean, contamination-free food and ensuring consumer acceptance of food with improved functional properties. Aspects of nanotechnology's application in relation to improving food nutrition and organoleptic properties have also been briefly addressed, as well as some insights into safety problems and regulatory concerns about nano-processed food products.

Key words: Nanoparticles, Food preservation, Functional food, Food safety, Food nutrition, Nano-processed food products

INTRODUCTION

The concept of nanotechnology was first discussed in 1959 by Richard Feynman in his talk "there's plenty of Room at the bottom". In 1974 Norio Taniguchi used the term nanotechnology for the first time. Nanotechnology produces things at molecular level from nanoscale components. Nano word derived from Greek root dwarf and Technology means to visualize, characterize, produce and manipulate matter of size of 1-100 nm. In layman's language, nanotechnology is the art and science of altering matter at the nanoscale or it is the study of making small microscopic things. Use of such kind of technique have raised with enhancing requirement of nanoparticle utilization in different fields food science and microbiology. Over the past few decades, nanotechnology has been positively regarded as attractive technology that has modernized the foodstuff division. It is a technology on the nanometer scale and deals with the atoms, molecules, or the macromolecules with the size of approximately 1–100 nm to create and use materials that have novel properties. Nanomaterials are the particles (crystalline or amorphous) of Organic or Inorganic materials having sizes in the range of 1-100 nm. The created nanomaterials possess one or more external dimensions, or an internal structure, on the scale from 1 to 100 nm that allowed the observation and manipulation of matter at the nanoscale(Gupta *et al.*, 2016). Nanotechnology provides a broad range of chances for the development and application of structures, materials, or system with new properties in various areas like agriculture, food, and medicine, etc (Roselli *et al.*, 2003). It brings about a great deviation in the food safety and quality. According to Dasgupta *et al.*, 2015, such nano-based techniques provides us overall food solutions from food manufacturing, processing to packaging. The use of Nano techniques in dietdivision can be concise in two main parts that are diet based nanostructured constituents and food nanosensing, can be utilized to enhance food safety and quality(Ezhilarasi *et al.*, 2013).

SIGNIFICANCE OF NANO-TECHNIQUES IN FOOD TREATING

The nanostructured food constitents have been enhanced that claim to provide better food taste and quality demensions(Cientifica Report, 2006). Nanotechniques assists to raise the storage span of numerous varieties of foodstuffs and also reduce the extent of wastage of food due to microbial attack(Pradhan *et al.*, 2015). Without disturbing their basic morphology of foodstuff, these days, nanocarriers are being used as delivery systems to carry food additives in food products(Ezhilarasi *et al.*, 2013). Nano polymers are trying to switch conventional materials in food packaging(Bratovˇci´c, 2015). Nanoparticles have better properties for encapsulation and release efficiency than traditional encapsulation systems, to enhance preservation and bioavailability of the active foodstuffs, numerous natural and man-made polymer based encapsulating transfer systems have been expounded in **Table 1**).

Significance of Nano-techniques in food treating can be examined by seeing its part in the food stuff improvisation in terms of- (i) food texture, (ii) food appearance, (iii) food taste (iv) nutritional value of the food, (v) food shelf-life.

TABLE 1-Distinct Nano-techniques to encapsulate and delivery of functional ingredients.

Nanotechnique	Feature	Examples	Reference
Edible coatings	• To preserve the quality of fresh foods during extended storage.	• Gelatin-based edible coatings containing celluloseNanocrystal • Chitosan/nanosilica coatings • Chitosan film with nano-SiO2 • Alginate/lysozyme nanolaminate coatings	Fakhouri *et al.*, 2014 Shi *et al.*, 2013 Yu *et al.*, 2012 Medeiros *et al.*, 2014
Hydrogels	• Can be easily placed into capsules, protects drugs from extreme environments, and to deliver them in response to environmental stimuli such as pH and temperature	• Protein hydrogels	Qui and Park, 2001
Polymeric micelles	• Solubilize water-insoluble compounds in thehydrophobic interior, high solubility, low toxicity	• PEO-b-PCL [poly (ethylene glycol) block-poly(caprolactone)] polymeric micelles • Methoxy poly(ethylene glycol) palmitate polymeric micelles	Ma *et al.*, 2008 Sahu *et al.*, 2008
Nanoemulsions	(i) Greater stability to droplet aggregation andgravitational separation; Beta-Carotene-based nanoemulsion (ii) Higher optical clarity (iii) increased oral bioavailability	• Beta-Carotene-based nanoemulsion	Kong *et al.*, 2011
Liposomes	• Since liposome surrounds an aqueous solution inside a hydrophobic membrane, it can be used delivery vehicles for hydrophobic molecules (contained within the bilayer) or hydrophilic molecules (contained in the aqueousinterior)	• Cationic lipid incorporated liposomes modified with an acid-labile polymer hyper-branched poly(glycidol) (HPG)	Yoshizaki *et al.*, 2014
Inorganic NPs	• They display good encapsulation capability and them rigid surfaces allow controlled functionalization	• Mesoporous silica nanoparticles	Tang *et al.*, 2012

NEED FOR FOOD NANO-TECHNOLOGY

1. Allows the placement of small structures placed with precision, simplicity and low cost.
2. Leads to economic growth.
3. Enhances national food security.
4. Improves Food quality.
5. Leads to job creation

Nutritional Value

A majority of bioactive compounds such as lipids, proteins, vitamins, lipids and carbohydrates, are sensitive to high acidic environment and enzyme activity of the stomach and duodenum.

(Koo *et al.*, 2005). The nanocomposite, nano-emulsification, and nanostructuration are the distinct technologies which have been used to encapsulate the substances in miniature forms to more efficiently transfer nutrients like protein and antioxidants for precisely targeted nutritional and health benefits. Polymeric nanoparticles are found to be suitable for the encapsulation of bioactive compounds (e.g., flavonoids and vitamins) to protect and transport bioactive compounds to target functions (Langer and Peppas, 2003).

NANO-TECHNIQUES IN FOODSTUFF WRAPPING

An ideal wrapping material must have moisture and gas absorbencypooled with biodegradable strength (Couch *et al.*, 2016). Smart nano-based technique offers a better wrapping material over conventional foodstuff wrapping technique from providing mechanical strength and antimicrobial films(Mihindukulasuriya and Lim, 2014). Many nanoparticles such as silver, copper, chitosan, and metal oxide nanoparticles like titanium oxide or zinc oxide have been reported to have antibacterial property (Tan*et al.*, 2013). Use of inert nanoscale fillers such as clay and silicate nano-platelets, silica (SiO2) nanoparticles, chitin or chitosan into the polymer matrix renders it lighter, stronger, fire resistance, and better thermal properties (Othman, 2014).

Many inorganic or organic fillers are being used in order toachieve improved polymer composites. The purpose of creating polymer composites is to have more mechanical and thermostable packing materials. The incorporation of nanoparticles in polymers has allowed developing more resist packaging material with cost effectiveness (Sorrentino *et al.*,2007).

NANO-BIOSENSORS FOR PATOHGEN INDENTIFICATION

In food microbiology, nano-biosensors are utilized for the identification of pathogens treating plants of foodstuffs (Cheng *et al.*,2006). They work as an indicator that responds to changes in environmental conditions such as humidity or temperature in storage rooms, microbial contamination, or products degradation (Bouwmeester *et al.*, 2009). Ex- thin films, nanorods and nano-fibres etc. (Jianrong *et al.*, 2004).

PROPERTIES OF AN IDEAL NANO DELIVERY SYSTEM

An ideal delivery system is supposed to have following properties:

(i) It must be able to transfer the active compound precisely at the target site

(ii) It must ensure accessibility at a target time and specific rate

(iii) effective to maintain active compounds at suitable levels for long periods of time (in storage condition).

Nanotechnology being used in the formation of encapsulation, emulsions, biopolymer matrices, simple solutions, and association colloids offers efficient delivery systems with all the above-mentioned qualities.

Fig. 1- Approaches of Nanotechnology

TOP-DOWN

- They arrange smaller components into more complex.
- Layer-by-layer self-assembly

BOTTOM-UP

- Create smaller objects by using larger ones.
- Utilizes principles of molecular recognition.

IMPLICATIONS OF NANOTECHNOLOGY

Health and Safety issues:

- Nanoparticles can cause serious illness or damage human body.
- Untraceable destructive weapons of mass destruction.

Social and Political issues

- Creates social strife through increasing wealth gap.
- Advisability of increasing scope of the technology creates political dilemma.

MERITS

1. By using Nano-biosensors, the presence of contaminants, mycotoxins, and microorganisms in food is reduced (Bratovˇci´c, 2015).
2. Nano based encapsulations hides odors, control interactions of active ingredients with the food matrix, control the release of the active agents, ensure availability at a target time and specificrate, and protect them from moisture, heat (Ubbink and Kruger,2006).
3. These nanoencapsulation transfer systems have ability to penetrate deeply into tissues due to their smaller size and thus allow efficient Transfer of active compounds to target sites in the body (Lamprecht *et al.*, 2004).
4. It is a matter of fact that surprisingly nanotechnology not only touches food texture, appearance, taste and storage life but has also brought about important changes in food products providing them new qualities.
5. Nanoparticles-based tiny comestible capsules with the target to enhance transfer of vitamins or breakable micronutrients in the regular foodstuff are being made to provide significant health boon (Koo*et al.*, 2005).
6. encapsulating the ingredients in minute forms to more efficient to transfernutrients like protein and antioxidants for accurately targeted nutritional and health boon. (Langer and Peppas, 2003).
7. The use of nanoparticles is not limited to antimicrobial food packaging but nanocomposite and nano laminates have been actively used in food packaging to provide a barrier from extreme thermal and mechanical shock enhance storage life(Sorrentino *et al.*, 2007).
8. The insertion of nanoparticles in polymers had provide more resistant cost effective wrapping material (Sorrentino et al., 2007).
9. Automatic pollution cleanup and Mass production in food and consumables.
10. Manufacturing at very low cost or no cost.

DEMERITS

1. Particle size directly affect the transfer of any bioactive compound to numerous sites within the body as it was noticed that in some cell lines, only submicron nanoparticles can be absorbed efficiently but not the larger size micro-particles (Ezhilarasi *et al.*, 2013).

2. Health and safety issues: Nano particles can cause serious illness or damage to human body. For ex- Carbon Nano-tubes can cause lung infection.
3. Loss of jobs in manufacturing and farming etc.
4. Creates social strife through increasing wealth gap.
5. Nano-pollution is created by toxic waste.
6. Atomic weapons could be more accessible and destructive.

Application of Nano-techniques in Food Industry

Nano-technology is among the most advanced fields of science and like every advanced field, it is full of both risks and opportunities. Several government and private bodies are investing in research on nanotechnology since the field holds immense potential. While the technology holds strong potential in various areas from medicine to IT and energy as well as environment, several risks associated with Nano technology have also been noted. Despite these risks, Nano technology remains a potentially attractive area and any major development in this field could become the next major breakthrough in the field of science and technology. Careful application of this technology could help the entire humanity.

- Food scientists say they have cooked up a way of using nanotechnology to make low-fat or fat-free foods just as appetizing and satisfying as their full-fat fellows.
- In medical field- end of illness (i.e. Cancer, heart diseases).
- Universal immunity- such diseases like AIDS and flu.
- Body sculpturing- changing of our appearance.

1. In Food treating

(a) **Anticaking agent-** Improves consistency and prevent lump formation.

(b) **Nano additives and Nutraceuticals-**Improve nutritional value of food.

(c) **Gelating agent-** Improve food texture.

(d) **Nano-carriers and Nano-capsulation-** To protect aroma, flavor and other food constituents.

2. In Food wrapping

(a) **Improved wrapping-** Utilization of nanoparticles to improve physical strength of food.

(b) **Active wrapping-** Utilization of nanoparticals as anti-microbial agent.

(c) **Smart wrapping-** Utilization of Nano-biosensors for pathogen identification. (Tan *et al.*, 2013).

REFERENCES

- Cientifica Report (2006). Nanotechnologies in the Food Industry, Published August 2006. Available at: http://www.cientifica.com/www/details.php?id47 [accessed October 24, 2006].
- Couch, L. M., Wien, M., Brown, J. L., and Davidson, P. (2016). Food nanotechnology: proposed uses, safety concerns and regulations. *Agro. Food Ind. Hitech*. **27**:36–39.
- Mihindukulasuriya, S. D. F., and Lim, L. T. (2014). Nanotechnology development in food packaging: a review. *Trends Food Sci. Technol*. **40:**149–167.
- Cheng, Q., Li, C., Pavlinek, V., Saha, P., and Wang, H. (2006). Surface-modified antibacterial TiO2/AgC nanoparticles: preparation and properties. *Appl. Surface Sci*. **252:**4154–4160.
- Bouwmeester, H., Dekkers, S., Noordam, M. Y., Hagens, W. I., Bulder, A. S., Heer, C., et al. (2009). Review of health safety aspects of nanotechnologies in food production. Reg. *Toxicol. Pharmacol*. **53:**52–62.
- Jianrong, C., Yuqing, M., Nongyue, H., Xiaohua, W., and Sijiao, L. (2004). Nanotechnology and biosensors. *Biotechnol. Adv*. **22:**505–518.
- Lamprecht, A., Saumet, J. L., Roux, J., and Benoit, J. P. (2004). Lipid nanocarriers as drug delivery system for ibuprofen in pain treatment. *Int. J. Pharma*. **278:**407–414.
- Fakhouri, F. M., Casari, A. C. A., Mariano, M., Yamashita, F., Mei, L. I., Soldi, V., et al. (2014). "Effect of a gelatin-based edible coating containing cellulose nanocrystals (CNC) on the quality and nutrient retention of fresh strawberries during storage," in Proceedings of the IOP Conference Series: Materials Science and Engineering, Conference 1, 2nd International Conference on Structural Nano Composites (NANOSTRUC 2014), Vol. 64, Madrid.
- Shi, S., Wang, W., Liu, L., Wu, S., Wei, Y., and Li, W. (2013). Effect of chitosan/nano-silica coating on the physicochemical characteristics of longanfruit under ambient temperature.*J. Food Eng*.**118:**125–131.
- Yu, Y., Zhang, S., Ren, Y., Li, H., Zhang, X., and Di, J. (2012). Jujube preservation using chitosan film with nano-silicon dioxide. *J. Food Eng*. **113:**408–414.
- Medeiros, B. G. D. S., Souza, M. P., Pinheiro, A. C., Bourbon, A. I., Cerqueira, M. A., Vicente, A. A., et al. (2014). Physical characterisation of an alginate/lysozymenano-laminate coating and its evaluation on 'Coalho' cheese shelf life. *Food Bioproc. Technol*. **7:**1088–1098.
- Qui, Y., and Park, K. (2001). Environment sensitive-hydrogels for drug delivery. *Adv. Drug Devl. Rev*. **53:**321–339.
- Ma, Z. S., Haddadi, A., Molavi, O., Lavasanifar, A., Lai, R., and Samuel, J. (2008). Micelles of poly (ethylene oxide)-b-poly(epsiloncaprolactone)

as vehicles for the solubilization, stabilization, and controlled delivery of curcumin. J. *Biomed.Mater. Res. A*. **86:**300–310.

- Pradhan, N., Singh, S., Ojha, N., Srivastava, A., Barla, A., Rai, V., et al. (2015). Facets of nanotechnology as seen in food processing, packaging, and preservation industry. BioMed Res. Int. 2015:365672.
- Ubbink, J., and Kruger, J. (2006). Physical approaches for the delivery of active ingredients in foods. *Trends Food Sci. Technol*. **17:**244–254.
- Sahu, A., Bora, U., Kasoju, N., and Goswami, P. (2008). Synthesis of novel biodegradable and selfassembling methoxy poly (ethylene glycol)-palmitate nanocarrier for curcumin delivery to cancer cells. *Acta Biomater*. **4:**1752–1761.
- Sorrentino, A., Gorrasi, G., and Vittoria, V. (2007). Potential perspectives of bionanocomposites forfood packaging applications. *Trends Food Sci. Technol.***18:**84–95.
- Kong, M., Chen, X. G., Kweon, D. K., and Park, H. J. (2011). Investigations on skin permeation of hyaluronic acid based nanoemulsion as transdermal carrier. Carbohydr. Polym. **86:**837–843.
- Koo, O. M., Rubinstein, I., and Onyuksel, H. (2005). Role of nanotechnology in targeted drug delivery and imaging: a concise review. Nanomed. *Nanotechnol. Biol. Med*. **1:**193–212.
- Langer, R., and Peppas, N. A. (2003). Advances in biomaterials, drug delivery, and bionanotechnology. AIChE J. **49:**2990–3006.
- Yoshizaki, Y., Yuba, E., Sakaguchi, N., Koiwai, K., Harada, A., and Kono, K. (2014). Potentiation of pH-sensitive polymer-modified liposomes with cationic lipid inclusion as antigen delivery carriers for cancer immunotherapy. Biomaterials**35:**8186–8196.
- Tang, F., Li, L., and Chen, D. (2012). Mesoporous silica nanoparticles: synthesis, biocom-patibility and drug delivery. Adv Mater. **24:**1504–1534.
- Tan, H., Ma, R., Lin, C., Liu, Z., and Tang, T. (2013). Quaternized chitosan as an antimicrobial agent: antimicrobial activity, mechanism of action and biomedical applications in orthopedics. *Int. J. Mol. Sci*. **14:**1854–1869.
- Bratovˇci´c, A., Odobaši´c, A., ´Cati´c, S., and Šestan, I. (2015). Application of polymer nanocomposite materials in food packaging. *Croat. J. Food Sci. Technol*. **7:**86–94.
- Gupta, A., Eral, H. B., Hatton, T. A., and Doyle, P. S. (2016). Nanoemulsions: formation, properties and applications. SoftMatter **12:**2826–2841.
- Roselli, M., Finamore, A., Garaguso, I., Britti, M. S., and Mengheri, E. (2003). Zinc oxide protects cultured enterocytes from the damage induced by Escherichia coli. *J. Nutri*. **133:**4077–4082.

- Dasgupta, N., Ranjan, S., Mundekkad, D., Ramalingam, C., Shanker, R., and Kumar, A. (2015). Nanotechnology in agrofood: from field to plate. *Food Res. Int*.**69:**381–400.
- Ezhilarasi, P. N., Karthik, P., Chhanwal, N., and Anandharamakrishnan, C. (2013). Nanoencapsulation techniques for food bioactive components: a review. *Food Bioprocess Technol*. **6:**628–647.
- Othman, S. H. (2014). Bio-nanocomposite materials for food packaging applications: types of biopolymer and nano-sized filler. Agric. *Agric. Sci. Proc*. **2:**296–303.
- https://notesmatic.com/2018/07/applications-advantages-and-disadvantages-of-nanotechnology/
- https://www.reuters.com/article/us-nano-food/scientists-see-risks-and-benefits-in-nano-foods-idUKTRE68E24W20100915

12

CHAPTER

CHALLENGES, TRENDS AND OPPORTUNITIES IN FOOD PROCESSING

Anjum Ayoub*[1], Kundan Veer Singh[2], Jagmohan Singh[1], Fozia Hameed[1] & Nadira Anjum[1]

[1]*Division of Food Science and Technology, Sher-e-Kashmir University of Agricultural sciences and Technoloy-Jammu, Chatha-180009, India*

[1]*Division of Plant Breeding and Genetics, Sher-e-Kashmir University of Agricultural sciences and Technoloy-Jammu, Chatha-180009, India*

*Corresponding Author: *anjumparay77@gmail.com*

ABSTRACT

This chapter addressed the problems that food processing has faced in the past, current, and future. New prospects for food production emerge if new problems can only be solved using new approaches. New trends, preservation, nutrition, sensorial, wellbeing, environmental, and future possibilities were among the topics covered by the challenges. Traditional, emerging, and re-emerging challenges were addressed in each issue, as well as new approaches, creative interpretation, and technologies that could be used to overcome them. Opportunities in Food Nanotechnology, probiotics and Prebiotics, additives, enzyme activity, nutrient retention, Functional foods. Also the various provocations faced by the food processing sector.

Keywords: Food processing, Opportunities, Past, Current and Future.

INTRODUCTION

These days' food processing has gained a level that is beyond the limit of imagination. Diverse kinds of food treating have useful or negative effects on food, diet quality, and human health. The food stuffs that we consumed today nearly all locations of the world is now processed in some way and they are practically omnipresent (Priyadarshini *et al*, 2019).To replace traditional food processing techniques, a new range of food treating technologies has been examined and advanced so that improved quality and more consumer liking foods stuff can be manufactured (Knoerzer *et al.*, 2011). According to Boekel *et al.*, 2010, Over the last decade, the main target has been on quality of food stuffs, raising its process of safety, efficiency, productivity and stability in a better way. The factors that prejudice nutrient quality of food stuffs are Whole food chain, from eminence of raw material, transportation, packaging, storage (Falguera *et al.*, 2012; Moskowitz *et al.*, 2012).

For people's consumption Numerous current methods and techniques involved in food treating are applied in order to transform raw material components into a better foodstuff. Food treating suggests the use of harvested, butchered and clean components to build food products for the market demand. (Moubarac *et al.*, 2014; Senorans *et al.*, 2003).These days, a striking shift process is taking place worldwide within alimentary production patterns and systems, with regard to the ready-for-consumption food, which replaces traditional food based on fresh meals. (Soon *et al.*, 2020).Researcher in this industry are currently working on various ways, certain processes such as drying, nonalcoholic fermenting, skimming, pasteurization, freezing, and vacuum-packing are rightly considered as beneficial in the food industry. However, other processes are regarded as less beneficial or even harmful for human health, namely, charring, hydrogenation, carbon dioxide addition, alcoholic fermenting, salt-pickling, and sugaring (Priyadarshini *et al*, 2019; Senorans *et al.*, 2003).At the present time, a radical variation has been noticed in how agreement and its penalties are loomed. at the local and state levels due to regulatory requirements, and to a much less tolerant and most knowledgeable public, the industry has been in the critical public eye and placed under a much more thorough scrutiny and examination than ever before, there has been an increasing demand of documentation, in terms of legislation. Legislation is seeing a lot of issues related to allergens, organics, and package labeling,it is very difficult to make a mistake and have no one find out about it. In the end, we are actually all consumers and expect safe food. (Nyarugwe *et al.*, 2019; Wenzl *et al.*, 2006; Eriksson *et al.*, 2020).

The Internet and social media have wedged the industry by letting an immediate exchange of information, which, at times, can include false or true data, hence rightly entitling food companies to have a congruently satisfactory reaction. With the altered way of foodstuff industry experts recognize food safety technology has greatly inhibited, react to, and communicate on specific food safety issues. (Ollinger *et al.*, 2011; Yeung R and Morris J, 2001). Services, processes, or innovations, understood as new products, are recognized as an important instrument for companies belonging to the food industry to stand out from competitors and to satisfy consumer expectations (MenradK, 2004; Samoggia A and Riedel B, 2019).Notwithstanding, nowadays, consumers' preferences for food have changed significantly; in fact, consumers believe that food should directly contribute to their health (Samoggia A and Riedel B, 2019; Black *et al.*, 2020). In these situations, the demand of consumers is no longer limited to satisfying hunger. Food should provide the necessary nutrients while at the same time preventing nutrition-related diseases and improving physical and mental well-being (Bowman BA and Mokdad AH, 2020). Recent food industry innovations mainly regard new technical and scientific approaches in food processing, as well as the implementation of new food products on the market. In this respect, due to the constantly rising cost of healthcare, the steady increase in life expectancy, and the necessity for the elderly to enjoy an improved life quality, functional food plays a major role in the consumers' demand nowadays. As such, researchers agree in stating that functional food represents one of the most rewarding and gratifying areas of research and innovation in the food industry (Annunziata A and Vecchio R., 2011; Granato *et al.*, 2020).The food industry is a key branch in the global economy. As a consequence, many authors highlighted its relevance for employment and economic output (Nyarugwe et al., 2019; Eriksson et al., 2020; MenradK, 2004).

New trends in food processing

Nowadays, a major trend in food manufacturing has been influenced by the consumers' demand of functional or health-promoting foods. According to the Food and Nutrition Board of the Conform Institute of Medicine, functional foods are defined as ''any food or food ingredient that may provide a health benefit beyond the traditional nutrients it contains. Functional food is the processed nutritious foods containing supplementary ingredients that give support to specific functions of the human body or according to Sloan, 2000, it is the food or beverage that imparts a physiological benefit that enhances overall health, helps prevent or treat a disease/ condition, or improves physical or mental performance via an added functional ingredient, processing modification, or biotechnology" (Priyadarshani *et al.* 2019; Senorans *et al.* 2003; Hasler CM, 1998; Sloan AE, 2000; Sloan AE, 2014).

Functional Foods

The major processes based on biotechnology are mainly used in the production of foods with specific functionalities. In this respect, genetic engineering constitutes an important support. Consequently, the industry's final objective of obtaining new foods of specific composition, with new and better functional properties, can be achieved, as these new biotechnologies can alter the genes contained in certain cells (Siro *et al.*, 2008; Bagchi D, 2019).

The raw materials with modified composition constitute another promising area for producing foods with enhanced nutritional value. For ex., foods having improved nutritional value began to appear 20 years ago, such as tomatoes with higher lycopene content or corn with higher oleic acid content (Pszczola DE,1998) and white wines with a higher concentration of resveratrol (Gonzalez *et al.*, 2000), hypoallergenic foods, in which a specific protein or peptide has been removed (Nakamura R and Matsuda T,1996).

Functional food products have also been produced by the wide use of enzymes. Hence, new advances in this area have been used such as techniques of enzyme and cell immobilization or new developments in bioreactors. For ex, milk protein concentrates with a decreased content of lactose (Szczodrak J,1999) and fats with a controlled content of fatty acids (Ki Teak L and Foglia TA, 2000).

Another important technique is the application of membrane technology, which is used in modifying the composition of foods and their functionality (Girard B and Fukumoto LR,2000). Mostly used foods in a liquid mediumeg- separation and concentration of milk components without the proteins, bioactive substances (Marella C,2013).

Various processes have been developed in the dairy industry, for instance, ultrafiltration (Mohammad *et al.*, 2012), nanofiltration (Salehi F,2014), electrodialysis (Mondor *et al.*, 2012), and lactoferrin(Wang *et al.*, 2019).

Opportunities in Food enriched with natural ingredients

The addition of Fe to bread or vitamin A and D to milk is an old and well known procedure called 'food fortification." The loss of nutrients while food is being processed has been overcome lately by supplementing health-promoting ingredients or nutraceutical ingredients to food products in this respect, consumers generally require to be offered all-natural foods, and in order to provide them with this type of products, natural ingredients have been manufactured by biotechnological, membrane technology, and supercritical fluid extraction (Bagchi D, 2019; Pszczola DE,1998).The most important elements used for food fortification are iron, vitamin A, iodine, folate (vitamin B9), Vitamin C, Vitamin D, calcium, selenium, fibres, proteins, and fatty acids (Vlaic *et al.*, 2019).

Opportunities to promote health -Probiotics and prebiotics

These are the part of so-called food that promote health, considered to be major nutrients that influence the physiology and gastrointestinal function (Salminen S., 1999).According to FAO/WHO definition, probiotics are defined as live microorganisms which when administered in adequate amounts confer a health benefit on 'live microorangism which when administered in adequate amounts confer a health benefit on the host. Probiotic have beneficial effects on health through mechanism of action, such as preventing colonization or pathogen adhesion, metabolism production, and immume system modulation by producing antibiodies to immunoglobulins (Chugh B and Kamal-Eldin A., 2020). Probiotics, according to Gipson, are " a substrate that is selectively used by host microorganism that confer a health benefit" (Gibson*et al.*, 2017).The most used probiotic compounds are inulin, fructo-oligosaccharides, fructans, lactulose, and galacto-oligosaccharides (Markowiak P and Slizewska K., 2017).

Low G.I Pasta

Currently the developing countries are facing the major challenges are to provide to more and more nutritious ready to eat foods for their busy consumer, to offer to their aging population a real opportunity to live, not only longer, but also healthier (Dutcosky *et al.*, 2006). Pasta a universal cosmopolitan starchy food with high acceptance rate, which acts as an excellent barrier for supplying nutrition for health cautious population around the world (Tudorica *et al.*, 2002). Pasta food industry is a rapid growing sector due to its cheaper, easy to cook, longer shelf life and it may be fortified or enriched conveniently. Its storage life depends upon its level of drying and processing conditions. (Kaur G, 2009). It may be easily added with other ingredients like colorants, flavors, proteins, dietary fiber supplements, vitamins and minerals etc. with objective to raise its nutraceutical features. Its nutritional advantages are fortified when specifically, distinct pulses as well as millets were added for making. This starchy food has been shown to have a low glycemic index (G.I), however there are many distinct varieties of pasta that are present throughout world. (Sicignano *et al.*, 2015). The Glycemic Index (GI) means increase of blood glucose capacity of carbohydrate foods. Numbers of low GI foods were very limited, and according to that we need a much wide range of products that will be required to a make a well balanced low-GI diet practicable. The three Low-GI food groups are- low-GI (less than 55), medium-GI (55-69) and high-GI (more than 70). Promoting small changes in blood glucose level after meal, can reduce the long term risk of type 2 diabetes mellitus and thereby it will be favorable for control and prevention of risk factor related to the metabolic and obesity like coronary heart diseases. In completely healthy pregnant women, diets containing low-GI can help to curb or reduce the risk of large gestational period age without rise in the number of small gestational age infants. In children and adults with diabetes, low-GI diets may improve glycosylated hemoglobin levels without raising LDL-C or the risk of hypoglycemia. So it's important to replace wheat flour with pulses flours, millets flour, and green leafy vegetables and to optimize processing conditions like particle size and water absorption. Processing of raw materials which have low-GI characteristics must be added (millet flour, pulses flour, spinach flour, Broccoli flour and Pomegranate peel flour) to prepare a new generation low-GI pasta. (Miller *et al.*, 2009; Silva *et al.*, 2013).

Pomegranate peel powder

They are the utmost wastage in food industry and they are mostly discarded and are also under-utilized but it contains an excellent source of bio-active compounds. Pomegranate peel powders are rich in flavonoids, anthocyanin, reservertol which are the richest source of antioxidants. Besides high antioxidant capacity, pomegranate peel extracts have been reported to possess a wide range of biological actions including microbial activity, anti-cancer activity, antidiarrheal activity, apoptotic and, anti-tyrosinase activity, anti-diabetic, and anti-inflamentary activities. Instead of discarding this waste it can be added to many food products as an ingredient. (Silva et al., 2013).

Okra seed oil

Okra (*Abelmoschusesculentus*(L.) Moench) is the most economically significant popular warm season vegetable, such as human diet, a good source of protein, carbohydrates, vitamins, Ca, K and minerals. It plays an important role for food security in developing countries. Okra seeds have been used on a small scale for oil production and its flour contains a good source of minerals, protein and fat which can help remove the stigma of food starvation (Williams JT, 2002; Gemede *et al.*, 2015; Farinde *et al.*, 2007; Hosain *et al.*, 2010). Being a multipurpose crop it is grown for its immature pods that can consumed as a fried or boiled vegetable or may be added to salads and soups (Gerrano AS., 2018).The demand for vegetable oils is rapidly increasing due to the growing human population, expanding oil industry with health-promoting oil components and the exploration of some underutilized and newer resources of vegetable oils. Seeds of lady's finger are light, yellowish like that of cottonseed and soybean oil. High saponification value and low acid values The demand for vegetable oils is rapidly increasing due to the growing human population, expanding oil industry with health-promoting oil components and the exploration of some underutilized and newer resources of vegetable oils. The low acid value The demand for vegetable oils is rapidly increasing due to the growing human population, expanding oil industry with health-promoting oil components and the exploration of some underutilized and newer resources of vegetable oils.(Yang *et al.*, 2014).The value of moisture in the seeds of lady's finger is very significant while preserving food. High moistures reduces the shelf life of foods by inviting various favorable microbes while low moisture keeps the spoilage at an insignificant level (Khan MR., 2012). Dry seed oil has a capability to be applied as a constituent in the manufacture of industrial soap, cosmetics, and for the thermal stabilizing polyvinyl chloride (PVC) (Akinhanmi *et al.* 2008). The bioavailability of okra good than other pulse and cereals crops. Okra oil is rich in essential fatty acids. It also exhibits good quality physico-chemical properties which are comparable with some conventional oils.

Provocations faced by Food processing sector

Distinct provocations faced by food processing sectors across the global years. (Aguilera JM 2018;Silva et al., 2018; Saguy et al., 2013). Silva et al. 2013, demonstrated an historic scenario of industrialization of food technology and its growth and development over the years. He described four different cycles related to food industries, between the 1950's to 1980's he described the first and second cycles that were linked with food product development and flavor enhancement. Between 1980's to 2000 he had focused on health parameter as a part of its 3rd cycle, while the current phase 4th cycle goes further and focused on wellbeing. Silva et al. 2018, also highlighted the present scenario and future challenge for the food industry to promote' 'the health of society' by promoting food ethics and human concern. Saguy et al. 2013,discuss the provocations faced by the World population growth and food security.

The first provocation that the humanity needs to solve safety, i.e., guarantee the food is safe to eat, although the emergence of microbiology process on early 1900's apparently guaranteed safety in food but still this a very current challenge (Simth JL and Fratamico PM, 2018). During 2018, The Rapid Alert System for Food and Feed of European Union (European Union annual report, 2018) identified 59 notifications of foodborne outbreak, being 20% of them identified with norovirus. Moreover, an outbreak of hepatitis A virus was linked to the frozen strawberries, same report also shows the presence of mycotoxins. The next provocation after 'safety' is 'food preservation' which needs to meet the consumer demand is the necessity to guarantee microbial chemical, physical and biochemical stability to safe products (Aguilera JM, 2019; Miller et al., 2011). The healthynatural isotonic coconut water whose flavor is damaged by thermal processing. High pressure processing has been already being used commercially to achieve the basic required safety, but unfortunately, this technology unable to inactivate polyphenol oxidase and peroxidase, resulting in pink color and impairing preservation. However, different non-conventional technologies were proposed as a possible alternative to inactivate these enzymes resistant to high pressure processing, such as dielectric barrier discharge plasma (Chutia et al., 2019) and photo-processing (Augusto *et al.*, 2015). Therefore, Combination of traditional and non-traditional technologies will be an interesting approach to enhance food processing. The third provocation is the nutritional quality, a nutrient which is present in food products in adequate quantities, but how our body will absorb and use that nutrient can vary widely, can be accelerated by food fortification (Rojas *et al.*, 2019; Verkempinck*et al.*, 2020). According to van Boekel *et al.*2020, food processing is still needs to be designed through a co-optimization between safety and quality, evaluating the safety factors generally applied with a scientific criticism. The fourth provocation is related with sensorial aspects. Even though safe and stable products are obtained with minimum nutrient degradation, aspects such as appearance, flavor, odour, taste and texture are sometimes neglected. Eg., ultra sound increases the juice consistency and reduce the pulp sedimentation and interesting results can have achieved by combination of technologies like in case rehydrated and thermal processed peas in comparison with natural vegetable. (Rojas et al., 2016).Moreover, there is increase in demand for food products with different sensations like texture and its relation with elderly people, exotic flavorsand even different colors. (Aquilera JM and Park DJ, 2016). Food is the central part of the lifestyle, it should contribute to consumer's perception, feeling (associated with consumer relations with foods based and their knowledge), satisfaction and happiness.

In fact, both nutritional quality and sensorial aspects must lie together (Aguilera JM, 2018).Another provocation which comes under the fifth number is the well-being for eg., number fexi-vegansis increasing (Gravel A and Doyen A, 2020), rising behaviour of consumers against processed foods (Verkempinck *et al.*, 2020). Food processing can produce more nutritive products by changing the nutrient bio-accessibility. According to Silva *et al.* 2018; Saguy IS., 2016 and Saguy et al. 2013 Firstly, we must teach population about what is food processing, a healthier diet and how to choose the best choices for each context of their lives. Secondly, we must search for alternatives to present food products whose composition and presentation make population more comfortable with them. The final provocation is the relation about the food by product processing and the acclimatization of the environment, Food by product industry consumes a relevant quantity of energy and virtual water pressure is still to reduce at beginning. Moreover, suitable part of energy is based on exhaustible sources (typically petroleum derivatives burned to produce vapour). simultaneously moving process towards less energy consumption and non-exhaustible sources is needed for a period, microwave and ohmic heating can be excellent alternatives for heating products with a better energy exchange and based on different exhaustible sources (solar and wind power for ex.,). However, microwaves or ohmic heating can substitute the pure steam injections in UHT process, also reduce the amount of water needed and simple fine the process (in special because vapour must be food grade). The production and consumption of insects in addition of cultural animal proteins are aslo reclaimed as a needed choice by requiring small use of water and land use, as well as remitting low levels of GHG's (greenhouse gases). (Mishyna*et al.*, 2020).

Opportunities in Food Additives

Food additives are nothing but the substances that are added to food to maintain or improve the appearance, freshness, taste safety, freshness, taste, texture of food. Some of them have been in use for centuries for preservation – such as salt (in meats such as bacon or dried fish), sugar (in marmalade). Various distinct food additives have been developed over time to meet the needs of food production, as making food on a large scale is very different from making them on a small scale at home. It serves as a well-defined technological function, such as to preserve the nutritional quality of the food or enhance the stability of the food. Additives can be extracted(from plants, animals, or minerals), or they could be artificially made. There are numerous thousand additives used in food, which are specifically designed to do a job in making food safer and more tempting. Based upon function, WHO together with FAO, classified food additives into 3 broad categories viz., Flavouring agents, Enzyme preparations and other additives.

1. Flavouring agents

They are incorporated into food to improve aroma or taste – make up the maximum number of additives utilized in foods. There are hundreds offlavourings used in a wide variety of foods, from confectionery and soft drinks to cereal, cake, and yoghurt. Natural flavouring agents include nut, fruit and spice blends, as well as those derived from vegetables and wine.

2. Enzyme preparations

They are naturally-occurring proteins that increase biochemical reactions by breaking down larger molecules into their smaller building blocks. These are the type of food additive that may or may not end up in the final food product. They can be obtained by extraction from plants or animal products or from micro-organisms such as bacteria and are used as alternatives to chemical-based technology. They are primarily utilized in baking (to improve the dough), for manufacturing fruit juices (to increase yields), in wine making and brewing (to improve fermentation), as well as in cheese manufacturing (to improve curd formation).

3. Other additives

They are chiefly utilized for preservation, coloring, and sweetening. They are incorporated when food is ready, parceled, transported, or warehoused. Preservatives can sluggish decomposition caused by air, bacteria, yeast or mould. Preservatives govern contamination that can cause foodborne illness, including life-threatening botulism. Colouring is added to food to replace colours lost during preparation, or to make food look more attractive. Non-sugar sweeteners are often used as an alternative to sugar because they contribute fewer or no calories when added to food. (The World Health Organization report, 2018).

Opportunities in Food Nanotechnology

Over the past a decade, Nanotechnology agreements whole nutrition solutions from food manufacturing, processing to packaging. Nanomaterials bring about a great variation not only in the foodstuff safety and quality but also in health benefits. Many organizations, researchers, and industries are coming up with advanced techniques, methods, and products that have a direct application of nanotechnology in food science (Dasgupta *et al.*, 2015). In food processing, theses nanostructures can be utilized as carriers for smart delivery of nutrients, anti-caking agents, antimicrobial agents, fillers for improving mechanical strength and durability of the packaging material, etc. whereas food nano-sensing can be applied to achieve better food quality and safety evaluation (Ezhilarasi *et al.*, 2013). The nanostructured food components help to enhance taste, texture, and consistency (Cientifica Report, 2006). Nanotechnology help to enhance the self-life and various types of foodstuffs that finally brought down the spoilage and wastage of food due to microbial infestation (Pradhan *et al.*, 2015). Nowadays Nano carriers are being utilized as delivery systems to carry food additives in food products without disturbing their basic morphology. Particle size may directly affect the delivery of any bioactive compound to various sites within the body as it was noticed that in some cell lines, only submicron nanoparticles can be absorbed efficiently but not the larger size micro-particles (Ezhilarasi*et al.*, 2013).

REFERENCES

- Annunziata A, Vecchio R. Functional foods development in the European market: A consumer perspective. Journal of Functional Foods. 2011;3(3):223-228. DOI: 10.1016/j.jff.2011.03.011
- Aguilera JM: Food engineering into the XXI century. *AIChE J* 2018, 64:2-11
- Aguilera JM: The food matrix: implications in processing, nutrition and health. *Crit Rev Food Sci Nutr* 2019, 59:3612-3629

- Augusto PE, lbarz R. Garvin A. lbarz A: Peroxidase (POD) and polyphenol oxidase (PPO) photo-inactivation in a coconut water model solution using ultraviolet (UV). *Food Res Int* 2015, 74: 151-159.
- Aquilera JM, Park DJ: Texture-modified foods for the elderly: status, technology and opportunities. *Trends Food SciTechnol* 2016, 57: 156-164.
- Boekel V, Martinus VF, Pellegrini N, Stanton C, Scholz G, Lalljie S, et al. A review on the beneficial aspects of food processing. Molecular Nutrition & Food Research. 2010;54(9):1215-1247
- Bagchi D. Nutraceutical and Functional Food Regulations in the United States and around the World. 3rded: USA: Academic Press Elsevier; 2019. DOI: 10.1016/B978-0-12-373901-8.X0001-7
- Black MM, Delichatsios HK, Story MT, editors. Nutrition Education: Strategies for Improving Nutrition and Healthy Eating in Individuals and Communities. Switzerland/S. Karger AG., Basel: Nestle NutrInst Workshop Ser. Nestle Nutrition Institute; 2020. P.92. DOI:10.1159/000499552
- Bowman BA, Mokdad AH. Addresing nutrition and chronic disease: Past, present, and future research directions. Food and Nutrition Bulletin. 2020;41(1):3-7. DOI; 10.1177/0379572119893904
- Chugh B, Kamal-Eldin A. Bioactive compounds produced by probiotics in food products. Current Opinion in Food Science. 2020;1-5. DOI: 10.1016/j.cofs.2020.02.003
- Chutia H, Kalita D, Mahanta CL, Ojah N, Choudhury AJ: Kinetics of inactivation of peroxidase and polyphenol oxidase in tender coconut water by dielectric barrier discharge plasma. *LWT- Food SciTechnol*2019, 101:625-629.
- Dutcosky SD, Grossmann MV, Silva RS, Welsch AK. Combined sensory optimization of a prebiotic cereal product using multicomponent mixture experiments. Food Chem. 2006; 98:630-638.
- Eupropean Union: *The Rapid Alert System for Food and Feed-* 2018 Annual Report. Available at: 2019 https://ec.europa.eu/food/sites/food/files/safety/docs/rasff_annual_report_2018.pdf.
- Gravel A, Doyen A: The use of edible insect proteins in food: challenges and issues related to their functional properties. Innov Food SciEmergTechnol 2020, 59 102272.
- Miller FA, Ramos BF, Gil MM, Brandao TR, Teixeira P, Silva CL. Heat inactivation of Listeria innocua in broth and food products under non-isothermal conditions. *Food Control* 2011, 22:20-26.
- Miller JB, Price JM, Steinbeck K, Caterson I. Dietary glycemic index: health implications. J Am Coll Nutr. 2009; 28:446S-449S.

- Mishyna M, Chen J, Benjamin O: Sensory attributes of edible insects and insect-based foods-future outlooks for enhancing consumer appeal. *Trends Food SciTechnol* 2020, 95: 141-148.
- Rojas ML, Alvim ID, Augusto PED: Incorporation of microencapsulated hydrophilic and lipophilic nutrients into foods by using ultrasound as a pre-treatment for drying: a prospective study. *UltrasonSonochem* 2019, 54: 153-161.
- Rojas ML, Leite TS, Cristianini M, Alvim ID, Augusto PE: Peach juice processed by the ultrasound technology: changes in its microstructure improve its physical properties and stability. *Food Res Int* 2016, 32:22-33.
- Silva VLS, Sereno A, Sobral PJA: Food industry and processing technology: on time to harmonize technology and social drivers. *Food Eng Rev* 2018, 10:1-13.
- Saguy IS: Challenges and opportunities in food engineering: modeling, virtualization, open innovation and social responsibility. *J Food Eng* 2016, 176:2-8.
- Saguy IS, Singh RP, Johnson T, Fryer PJ, Sastry SK: Challeges facing food engineering. *J Food Eng* 2013, 119:332-342.
- Simth JL, Fratamico PM: Emerging and re-emerging foodborne pathogens. *Foodborne Pathog Dis* 2018, 15:737-757.
- Silva E, Birkenhake M, Scholten E, Sagis LM, Van-der-Linden E. Controlling rheology and structure of sweet potato starch noodles with high broccoli powder content by hydrocolloids. Food Hydrocoll. 2013; 30:42-52.
- Sicignano A, Di Monaco R, Masi P, Cavella S. From raw material to dish: pasta quality step by step. J Sci Food Agric. 2015; 95:2579-2587.
- van Boekel MAJS, terSteeg PF, Dahoe AE: Co-optimization of safety, quality and legislation: opening Pandora's box?*CurrOpin Food Sci* 2020 http://dx.doi.org/10.1016/j.cofs.2020.02.001.
- Verkempinck SHE, Pallares AP, Hendrickx ME, Grauwet T: Processing as a tool to manage digestive barriers in plant based foods: recent advances. *CurrOpin Food Sci* 2020 http://dx.doi.org/10.1016/j.cofs.2019.11.007
- Kaur G. Development of technology for dietectic pasta. M.Sc. thesis, Punjab Agricultural University, Ludhiana, India. 2009.
- Tudorica CM, Kuri V, Brennan CS. Nutritional and physicochemical characteristics of dietary fiber enriched pasta. J Agric Food Chem. 2002; 50:347-356.
- Priyadarshini A, Rajauria G, O'Donnell CP, Tiwari BK. Emerging food processing technologies and factors impacting their industrial adoption. Critical Reviews in Food Science and Nutrition. 2019;59(19):3082-3101. DOI: 10.1080/10408398.2018.1483890

- Knoerzer K, Juliano P, Roupas P, Versteeg C. Innovative Food Processing Technologies: Advances in Multiphysics Simulation. Chichester, West Sussex, UK: John Wiley & Sons and Institute of Food Technologists; 2011. DOI: 10.1002/9780470959435
- Falguera V, Aliguer N, Falguera M. An integrated approach to current trends in food consumption: Moving toward functional and organic products? Food Control. 2012;26(2):274-281. DOI: 10.1016/j.foodcont.2012.01.051
- Moubarac J-C, Parra DC, Cannon G, Monteiro CA. Food classification systems based on food processing: Significance and implications for policies and actions: A systematic literature review and assessment. Current Obesity Reports. 2014;3(2):256-272. DOI:10.1007/s13679-014-0092-0
- Senorans FJ, Ibanez E, Cifuentes A. New trends in food processing. Critical Reviews in Food Science and Nutrition. 2003;43(5):507-526. DOI: 10.1080/10408690390246341
- Soon JM, Brazier AKM, Wallace CA. Determining common contributory factors in food safety incidents-A review of global outbreaks and recalls 2008-2018. Trends in Food Science & Technology. 2020; 97:76-87. DOI: 10.1016/j.tifs.2019.12.030
- Nyarugwe SP, Linnemann AR, Luning PA. Prevailing food safety culture in companies operating in a transition Economy-Does product riskiness matter? Food Control. 2019; 107:106803. DOI: 10.1016/j.foodcont.2019.106803
- Wenzl T, Simon R, Anklam E, Kleiner J. Analytical methods for polycyclic aromatic hydrocarbons (PAHs) in food and the environment needed for new food legislation in the Eupropean Union. TrAC Trends in Analytical Chemistry. 2006;25(7):716-725. DOI: 10.1016/j.trac.2006.05.010
- Eriksson D, Custers R, Bjornberg KE, Hansson SO, Purnhagen K, Qaim M, et al. Visser options to reform the European union legislation on GMOs: Scope and definitions. Trends in Biotechnology. 2020;38(3):231-234
- Ollinger ME, Muth MK, Karns S, Choice Z. Food safety audits, plant characteristics, and food safety technology use in meat and poultry plants. USDA-ERS Economic Information Bulletin. 2011; 82:1-44. DOI:10.2139/ssrn.2131423
- Yeung R, Morris J. Food safety risk: Consumer perception and purchase behavior. British Food Journal. 2001;103(3):170-187. DOI: 10.1108/000700110386728
- Menrad K. Innovations in the food industry in Germany. Research Policy. 2004;33(6-7):845-878. DOI: 10.1016/j.respol.2004.01.012

- Samoggia A, Riedel B. Assesment of nutrition-focused mobile apps' influence on consumers' healthy food behaviour and nutrition knowledge. Food Research International. 2019; 108766:1-30. DOI: 10.1016/j.foodres.2019.108766
- Granato D, Barba FJ, Kovacevic DB, Lorenzo JM, Cruz AG, Putnik P. Functional foods: Product development, technological trends, efficacy testing, and safety. Annual Review of Food Science and Technology. 2020;11(1):93-118. DOI:10.1146/annurev-food-032519-051708
- Hasler CM. Functional foods: Their role in disease prevention and health promotion. Food Technology. 1998; 52:63-70
- Sloan AE. The top ten functional food trends. Food Technology. 2000; 54:33-62
- Sloan AE. The top ten functional food trends. Food Technology (Chicago). 2014;68(4):22-45
- Siro I, Kapolna E, Kapolna B, Lugasi A. Functional food. Product development, marketing and consumer acceptance-A review. Appetite. 2008;51(3):456-467. DOI: 10.1016/j.appet.2008.05.060
- Pszczola DE. The ABCs of nutraceutical ingredients. Food Technology. 1998; 52:30-37
- Gonzalez Candelas L, Gil JV, LamuelaRaventos RM, Ramon D. The use of transgenic yeasts expressing a gene encoding a glycosyl hydrolase as a tool to increase resveratrol content in wine. International Journal of Food Microbiology. 2000; 59:179-183
- Nakamura R, Matsuda T. Rice allergenic protein and molecular-genetic approach for hypoallergenic rice. Bioscience, Biotechnology, and Biochemistry. 1996; 60:1215-1221
- Szczodrak J. Hydrolysis of lactose in whey permeate by immobilized beta-galactosidase from Penicilliumnotatum. ActaBiotechnologica. 1999; 19:235-250
- Ki Teak L, Foglia TA. Synthesis, purification, and characterization of structured lipids produced from chicken fat. Journal of the American Oil Chemists' Society. 2000; 77:1027-1034
- Girard B, Fukumoto LR. Membrane processing of fruit juices and beverages: A review. Critical Reviews in Food Science and Nutrition. 2000;40(2):91-157. DOI: 10.1080/10408690091189293
- Marella C. Application of membrane separation technology for developing novel dairy food ingrediants. Journal of Food Processing & Technology. 2013;04(09):1-5. DOI: 10.4172/2157-7110.1000269
- Mohammad AW, Ng CY, Lim YP, Ng GH. Ultrafiltration in food processing industry: Review on application, membrane fouling, and fouling control.

Food and Bioprocess Technology. 2012;5(4):1143-1156. DOI: 10.1007/s11947-012-0806-9

- Salehi F. Current and future application for nanofiltration technology in the food processing. Food and Bioproducts Processing. 2014;92(2):161-177. DOI: 10.1016/j.fbp.2013.09.005
- Mondor M, Ippersiel D, Lamarche F. Electrodialysis in food processing. In: Boye J, Arcand Y, editors. Green Technologies in Food Production and Processing. Food Engineering Series. Boston, MA: Springer; 2012. ISBN: 978-1-4614-1586-2
- Wang B, Timilsena YP, Blanch E, Adhikari B. Lactoferrin: Structure, function, denaturation and digestion. Critical Reviews in Food Science and Nutrition. 2019;59(4):580-596. DOI: 10.1080/10408398.2017.1381583
- Vlaic RA, Muresen CC, Muste S, Muresan A, Muresan V, Suharoschi R, et al. Food Fortification through Innovative Technologies. London, UK: EdituraIntechOpen; 2019. DOI: 10.5772/intechopen.82249
- Salminen S. Probiotics: Scientific support for use. Food Technology. 1999; 53:66
- Gibson GR, Hutkins R, Sanders ME, Prescott SL, Reimer RA, Salminen SJ, et al. Expert consensus document: The International Scientific Association for Probiotics and Prebiotics (ISAPP) consensus statement on the definition and scope of prebiotics. Nature reviews. Gastroenterology &Hepatology. 2017; 14:491-502
- Markowiak P, Slizewska K. Effects of probiotics, prebiotics, and synbiotics on human health. Nutrients. 2017; 9:1021
- Yang L, Takase M, Zhao T, Wu X. Potential non-edible oil feedstock for biodiesel production in Africa: a survey. Renew Sustain Energy Rev. 2014; 38:461-477.
- Gemede HF, Ratta N, Haki GD, Woldegiorgis AZ, Beyene F. Nutritional quality and health benefits of Okra (Abelmoschusesculentus): A review. J Food Process Technol. 2015; 6:2.
- Farinde AJ, Owolarafe OK, Ogungbemi OI. An overview of production, processing, marketing and utilization of Okra in egbedore local government area of Osun State, Nigeria. Agricultural Engineering International: CIGR Journal. 2007; 9:1-17.
- Hosain MM, Jannat R, Islam MM, Sarker MK. Processing and preservation of Okra pickle. Progressive Ageric. 2010; 21:215-222.
- Gerrano AS. Agronomic performance, nutritional phenotyping and trait associations of Okra (*Abelmoschusesculentus*) Genotypes in South Africa. Rediscovery of Landraces as a Resource for the Future. 2018:69.

- Badrle N. Nutrient profile, bioactive components, and functional properties of Okra (*Abelmoschusesculentus*(L.) Moench). Infruits, Vegetables, and Herbs 2016:365-409. Academic Press.
- Williams JT. Global research on underutilized crops: An assessment of current activities and proposals for enhanced cooperation. BioversityInternational. 2002.
- Khan MR. Osmotic dehydration technique for fruits preservation. A review. Pak J Food Sci. 2012; 22:71-85.
- Akinhanmi TF, Atasie VN, Akintokun PO. Chemical composition and physiochemical properties of cashew nut (Anacardiumoccidentale) oil and cashew nut shell liquid. J Agric, Food Environ Sci. 2008;2:1.
- Dasgupta, N., Ranjan, S., Mundekkad, D., Ramalingam, C., Shanker, R., and Kumar, A. (2015). Nanotechnology in agrofood: from field to plate. Food Res. Int. 69, 381–400. doi: 10.1016/j.foodres.2015.01.005
- Ezhilarasi, P. N., Karthik, P., Chhanwal, N., and Anandharamakrishnan, C. (2013). Nanoencapsulation techniques for food bioactive components: a review. Food Bioprocess Technol. 6, 628–647. doi: 10.1007/s11947-012-0944-0
- Cientifica Report (2006). Nanotechnologies in the Food Industry, Published August 2006. Available at: http://www.cientifica.com/www/details.php?id47 [accessed October 24, 2006].
- Pradhan, N., Singh, S., Ojha, N., Srivastava, A., Barla, A., Rai, V., et al. (2015). Facets of nanotechnology as seen in food processing, packaging, and preservation industry. BioMed Res. Int. 2015:365672. doi: 10.1155/2015/365672
- Ref- https://www.who.int/news-room/fact-sheets/detail/food-additives

INDEX